Decision Tables

Wiley Communigraph Series on Business Data Processing

Richard G. Canning and J. Daniel Cougar, Editors

Personnel Implications for Business Data Processing
Robert A. Dickmann

Design of Sequential File Systems
Thomas R. Gildersleeve

Management Reporting Systems
James M. McKeever in collaboration with Benedict Kruse

A Guide to Packaged Systems
Robert V. Head

Control of the Information System Development Cycle
Robert I. Benjamin

Decision Tables: Theory and Practice
Solomon L. Pollack, Harry T. Hicks, Jr., and William J. Harrison

An Introduction to Data Base Design
John K. Lyon

Decision Tables:

Theory and Practice

Solomon L. Pollack
S. D. Leidesdorf & Company
New York, New York

Harry T. Hicks, Jr.
Information Management, Inc.
San Francisco, California

William J. Harrison
Fireman's Fund Insurance Company
San Francisco, California

Wiley-Interscience

a Division of John Wiley & Sons, Inc.
New York · London · Sydney · Toronto

Copyright © 1971, by John Wiley & Sons, Inc.

All rights reserved. Published simultaneously in Canada.

No part of this book may be reproduced by any means, nor transmitted, nor translated into a machine language without the written permission of the publisher.

Library of Congress Catalogue Card Number: 76-150612

ISBN 0-471-69150-X

Printed in the United States of America

10 9 8 7 6 5 4 3 2 1

To Ginny, Elaine, and Anna

(Understanding wives, all)

Foreword

Inventing a new methodology is an exciting, yet often frustrating, experience. The joy of creation shared by a number of people working together is hard to match, but the slow acceptance of the new concept leads one to doubt the value of the invention itself.

Sol Pollack and I, together with a small group of dedicated professionals, shared this combined joy/frustration in developing decision tables. Throughout 1963 we had the pleasure of creating, defining, and proselytizing. Since 1963 we have had the frustration of finding that business analysts, system engineers, and computer programmers did not immediately accept decision tables as a major technique to assist in program design, implementation, and maintenance.

One of the most serious barriers to getting decision tables accepted as a standard technique has been the lack of a book which logically and fully covered the conceptual, mathematical, procedural, and operational foundations of decision tables. The coauthors have taken the time necessary to do a thorough, competent, professional job in all of these areas. I believe this is a comprehensive book which may be used by students as well as professional and managerial personnel to gain an understanding of the techniques and use of decision tables.

Decision tables provide a means for organizing your thoughts, clarifying the conditions behind alternate actions, and communicating this logic to someone else. By its highly structured nature it permits various explicit techniques to be used to ensure accuracy, eliminate redundancy, and improve programmability. The original purpose of decision tables was for use in preparing information for computer programmers. I find today, however, that much of my own use relates to a comprehensive examination of business and technical alternatives; that is, to understand and explain the various conditions under which I wish to take different courses of action. It has been an invaluable technique to get a number of people to focus on a single problem, agree on the criteria, state the alternatives available, and finally accept the relation between the criteria and the alternatives.

Today even those people most involved with the development of decision tables recognize that decision tables are not able to eliminate all the difficulties involved in handling complex decisions. Rather they are a most valuable addition to the set of tools that any good programmer, systems engineer, and businessman should have. They seem to be accepted readily by those who are comfortable with parallel thought processes. They appeal to the individual capable of examining ideas in parallel; the individual who does not require that each process be sequential. This capability requires training and, therefore, the individual must be willing to dedicate a certain amount of time to skill development.

The payoff, however, is very substantial. When you are skilled in using decision tables you may be sure that when you provide a solution statement to a problem it is accurate, complete, and capable of being used in a precise fashion.

So I wish all of you the pleasure of working with the technique of decision tables. I am confident you will enjoy learning how to use it from this fine book.

B. Grad

Series Preface

The Wiley Communigraph Series on Business Data Processing is intended for professionals and for persons desiring to improve their competence in business management applications of the computer. Within each of some twenty subject areas, publications will be provided over a range of technical depth. The objective is a series of publications that enable readers to gain an understanding in specialized subject areas.

The term "communigraph" was coined to reflect this philosophy—succinct treatment of specialized subjects. A communigraph requires only a modest investment in time on the part of the reader. The cummunigraph format has also allowed many authorities in the computer field—persons who have too many demands on their time to allow them to write full-length books—to participate as authors.

The series is designed to cover three levels of interest for detailed technical information:

Red Titles. For data processing managers, business programmers, analysts, and others who wish to gain background information about this subject.

Blue Titles. For business programmers and analysts, and data processing managers, who have some experience in this subject and who wish to study it in more technical depth.

Gold Titles. For senior business programmers and analysts, and data processing managers, with extensive experience in the subject, who wish a highly technical discussion of it.

The increase in breadth, complexity, and the dynamic characteristics of the computer field make a series such as this imperative—for training and for reference. The breadth is illustrated by the many subject areas covered in the series: from data base design to business planning by way of simulation; from hardware-software evaluation and selection to personnel considerations for business data processing. The growth of complexity is apparent in each of these subject areas; for instance, the variations in data base design and data management systems are increasing almost weekly. The dynamics of the field are obvious, with the multitude of new hardware and software products reaching the market each year.

In the face of this growth in depth and complexity, practitioners encounter the problems of developing and maintaining technical competence. The novice must upgrade his competence to reach the experienced practitioner level. The experienced system analyst and programmer must strive to reach senior levels. The professional must broaden his knowledge to include new specialties that adjoin his areas of technical competence.

Training, updating, and upgrading professionals in the computer field will be a continuous problem. We hope that the communigraph series will be an effective method in helping to cope with this problem.

RICHARD G. CANNING
J. DANIEL COUGER
Series Editors

Contents

Part I	INTRODUCTION	1
Section 1	Historical Discussion	3
Section 2	Decision Table Structure	7
Part II	THEORETICAL FOUNDATIONS	11
Section 3	Development of Decision Table Theory	13
Section 4	Decision Table Theorems	21
Part III	ANALYSIS AND IMPLICATIONS OF DECISION TABLES	25
Section 5	Conditions: Criteria for Decision	27
Section 6	Actions	39
Section 7	Optimization of Decision Tables	45
Section 8	The Rule: Redundancy and Contradiction	53
Section 9	The Rule: Completeness and Combination	55
Section 10	The Languages Used in Tables	63
Part IV	USES AND DEVELOPMENT OF DECISION TABLES	71
Section 11	Systems Analysis	73

Section 12	Decision Tables in Computer Programs	85
Section 13	Writing the Decision Table	93
Section 14	Debugging Programs Containing Decision Tables	97
Section 15	Maintenance of Decision Tables	107
Section 16	Decision Table Interactions	115
Section 17	A Case Study: Development and Decomposition of a Decision Table	119
Part V	TABLE TRANSLATION	127
Section 18	Decomposition Algorithms	129
Bibliography		139
Appendix I	Proof of Decision Table Theorems	145
Appendix II	An Introduction to Cobol	154
Appendix III	Decision Table Translation Algorithms	160
Index		175

Part I

INTRODUCTION

SECTION 1

Historical Discussion

Tables are a familiar part of everyone's life. From mathematical tables to the box score of yesterday's baseball game, they provide us with an orderly presentation of data. While such tables often assist us in making a decision, they are not decision tables. A decision table is a special form of table that codifies a set of decision rules based on a specific, clearly identified set of conditions and resulting actions. While the history of tables in general can be told in terms of centuries, decision tables are a fairly recent phenomenon.

In November 1957 General Electric initiated a research effort called the "Integrated Systems Project." Its objective was the study of the manufacturing processes that occur from the receipt of a customer order through the production of the finished product and the part that computers might play in them. It soon became apparent that the available methods of describing decisions—flowcharts, formuli, narratives, and the like—were inadequate for expressing the complex logic encountered in the processes being studied. For this reason the project team began a search for a new method of expression that culminated in the development of "decision structure tables" and a computerized method for solving them. These tables had all the characteristics of what we know today as decision tables but had a format similar to the truth tables from which they originated. Examples of a truth table and a TABSOL table are shown in Figures 1-1 and 1-2. The processor for solving these tables operated initially on an IBM 702 and was successively implemented on an IBM 305, 650, and 704. An improved processor and language called TABSOL were implemented on the GE 225 in early 1961.

At approximately the same time, and independently of GE's efforts, the Sutherland Company developed a decision table different in form but identical in concept. Whereas

4 HISTORICAL DISCUSSION

A	B	A ∨ B	A ∧ B
T	T	T	T
T	F	T	F
F	T	T	F
F	F	F	F

Figure 1-1 Truth Table indicates for each of the truth values that A and B can assume, the truth value of the logical statements "A OR B," "A AND B." The table is read horizontally.

General Electric developed the concept of decision tables and the computer-based solution method almost simultaneously, Sutherland developed their tables strictly as an aid to system analysis and documentation, leaving the solution of the table to the programmer. As a result, the subject matter expressed within the two forms of tables differed. General Electric utilized their tables to describe manufacturing decisions in great detail, whereas Sutherland used tables to express more general "management rules"—expressions of policy independent of the eventual processing media. As was the case with GE, Sutherland was forced to develop decision tables out of desperation. They had expended almost one man-year of effort attempting to specify the logic of a complex file update procedure without useful results. When the effort was restarted using decision tables, it was completed successfully in about 12 man-weeks. Following this initial experience, Sutherland continued using decision tables for documenting a wide variety of systems. An ex-

ITEM-1	ITEM-2 EQ	ITEM-3 EQ	GO TO
EQ 4	3	05	TABLE-2
EQ 6	4	10	TABLE-2
GR 7	5	15	TABLE-3

Figure 1-2 A decision structure table in **TABSOL** format. Decision rules are read horizontally; for example the first rule reads, "If **ITEM-1 EQ** 4 and **ITEM-2 EQ** 3 and **ITEM-3 EQ** 05, then **GO TO TABLE-2**."

		L-CODE = 6		
		DESC = 8	DESC = 9	DESC = 10
CLASS = 2	TYPE = 4	ACTION-1	ACTION-1	ACTION-2
	TYPE = 3	ACTION-3	ACTION-2	ACTION-1

Figure 1-3 An example of a decision table used for man-to-man communication. The conditions are listed both horizontally on the left and vertically across the top, with the actions shown in the center. Just as the **TABSOL** table is related to truth tables, this form appears related to the Karnaugh map.

ample of a communication-oriented decision table is shown in Figure 1-3.

Another early user, Hunt Foods and Industries, began using decision tables as an aid in man-to-man communication in 1959. This work was described in one of the earliest published works on decision tables.[1]

In May 1959 the Conference on Data Systems Languages (CODASYL, the organization that developed COBOL) was convened. It designated one of its committees, the Systems Group, to pursue the objective of developing a machine-independent, systems-oriented language. After reviewing several approaches to this objective, they began to study decision tables—an effort that was to occupy two years and result in a decision table language known as DETAB-X (Decision Tables, Experimental).

In 1960, General Electric presented their work on decision tables at the Eastern Joint Computer Conference and during the next two years a great deal of effort, most of it unreported in the literature, was spent in

---

[1] Orren Y. Evans, "Advanced Analysis Method for Integrated Electronic Data Processing." *IBM General Information Manual*, F20-8047.

developing decision table processors. IBM guided the implementation of at least three decision table processors, one on the IBM 1401, one on the 7080 in cooperation with the Boeing Company, and one on the 7090 in conjunction with RAND Corporation (FORTAB). Insurance Company of North America produced a decision table processor of their own called LOBOC, also on the IBM 7080, and GE, as noted earlier, implemented TABSOL on the GE-225.

In September 1962 the CODASYL Systems Group held a seminar in New York to present the results of their study of decision tables to the public. The product of their effort, called DETAB-X, consisted of a language supplement to COBOL-61 to be used within the framework of decision tables. The seminar featured talks by the early developers: Sutherland, General Electric, Insurance Company of North America, RAND, and IBM. Its objective was to stimulate interest and experimentation in the use of decision tables and their translators. In spite of the enthusiasm of the Systems Group and the information content of the seminar, the experimentation and resulting exchange of information that was hoped for never took place. The Group shortly moved on to other projects, leaving as their testament the format of decision table now accepted as "standard." An example of this format is shown in Figure 1-4.

The period from the DETAB-X seminar until 1965 was marked by inactivity. Few articles were written and little expansion in the use of decision tables by new users was observable. Then, in June 1965, the Special Interest Group for Programming Languages (SIGPLAN) of the Los Angeles Chapter of the Association for Computing Machinery appointed a working group to develop a decision table preprocessor. In order to guarantee a wide distribution, the preprocessor was written in a restricted subset of COBOL and accepted decision tables coded in COBOL to convert them to COBOL source code. The preprocessor, called DETAB/65, was released free of charge and distributed through the Joint Users Group. Although it was implemented on a number of computers, including the CDC 1604, 3400, and 3600 and the IBM 7040, 7044, and 7094, its admittedly inefficient conversion algorithm and lack of maintenance led to its disuse and eventual disappearance. It seems evident, however, that DETAB/65 was the ancestor of the current group of proprietary decision table preprocessors that have been developed since 1966. These preprocessors generally follow the DETAB/65 design—a preprocessor written in COBOL that converts decision tables containing COBOL components to a stream of COBOL code suitable for presentation to a compiler. The exception is IBM's System/360 Decision Logic Translator that processes decision tables coded in FORTRAN. Some of the preprocessors originally developed for COBOL offer the option of using FORTRAN.

Table 3	1	2	3	4	ELSE
FIELD-1 = 3	Y	Y	N	N	–
FIELD-2 =	3	4	10	15	–
FIELD-3 =	ZERO	ZERO	POSITIVE	NEGATIVE	–
MOVE A-6 TO A-7	X	–	X	X	–
GO TO	TABLE-4	TABLE-4	TABLE-5	TABLE-6	TABLE-9

Figure 1-4 A decision table in the "standard" format. Conditions and actions are listed on the left-hand side with decision rules read vertically from top to bottom; for example, Rule 2 reads "If **FIELD-1 = 3** and **FIELD-2 = 4** and **FIELD-3 = ZERO**, then **GO TO TABLE-4**."

In summary, the history of decision tables can be viewed as consisting of four eras:

1. The era of initial development, 1957–1960.
2. The first era of preprocessors, 1961–1962.
3. The era of silence, 1963–1965.
4. The second era of preprocessors, 1966–present.

A review of this history leads one to wonder why decision tables are not the universally used technique for systems analysis and program development. The literature of decision tables is replete with stories of how this technique succeeded where the more widely used methods of flow charting and narrative had failed. Indeed, the early "pioneers" were driven to invent decision tables because they could not solve their problems by any other means and many subsequent users first turned to decision tables for the same reason. This being true, why has the use of decision tables not been wider?

Three possible causes can be identified. First, there has been limited information on decision tables available to the main body of systems analysts and programmers. Although the bibliography of this book is not small by any means, many of the earlier articles represent highly technical discussions of conversion methods (which assume prior knowledge), or were published in proceedings, technical journals, or other media not generally read by the commercial practitioner. Computer manufacturers have not, as a rule, taught decision tables in their introductory programming courses, whereas they invariably teach flowcharting.

Second, the use of decision tables requires a different way of looking at problems than does flowcharting—the method of analysis learned first by analysts and programmers. Flowcharting leads one to adopt a sequential model of decision making—a test followed by one or more actions, then another test or two, and so on. Decision tables, on the other hand, require an overall analysis of the conditions that comprise a given problem and the effect of their various combinations on the solution. It is only natural that analysts and programmers, trained in a sequential type of analysis, resist shifting their fundamental outlook to accommodate decision tables.

Third, there has been a general lack of decision table translators available to the data processing community. Most of those developed before DETAB/65 were distributed in a limited way if at all. This meant that most decision tables that were developed had to be hand translated to sequential code for input to the computer. Experience has shown that the absence of a mechanized means of translation will result in a rapid decrease of interest in decision tables on the part of the programming staff.

These three conditions are now in the process of amelioration. This book is the sixth published in the last two years on the subject of decision tables. In addition, seminars conducted by both ACM and several private consultants have been and are currently increasing the number of analysts and programmers who know of decision tables. This increase of knowledge, coupled with the availability of a number of preprocessors will, one hopes, overcome the historical resistance to decision tables.

SECTION 2

Decision Table Structure

This chapter introduces the reader to decision tables by describing their structure. The theory, use, advance practice, and techniques of decision tables are fully discussed in the remainder of this book.

DECISION TABLE STRUCTURE

As a prelude to defining a decision table, we define a *decision rule* as a statement that prescribes the set of conditions that must be satisfied in order that a series of actions be executed; for example, the following is a decision rule:

If the employee is entitled to overtime pay and he worked more than 40 hours this week, pay him his regular wages plus the product of his overtime rate times his hours in excess of 40.

Other rules can be derived from the above two conditions: overtime eligibility and whether the employee has worked more than 40 hours in the week. The table that describes all of these possible decision rules is a decision table.

A *decision table* is a structure for describing a set of related decision rules. The basic parts of a decision table are shown in Figure 2-1. All boxes above the horizontal double line are condition boxes; those below are action boxes. Each box to the left of the vertical double line is called a *stub*, each box to the right is called an *entry*. The combination of a stub and an entry forms a condition (above the horizontal double line) or forms an action (below the horizontal double line). Each column to the right of the vertical double line is automatically combined with the one column to the left of the vertical double line to form a decision rule. The topmost horizontal line repre-

8 DECISION TABLE STRUCTURE

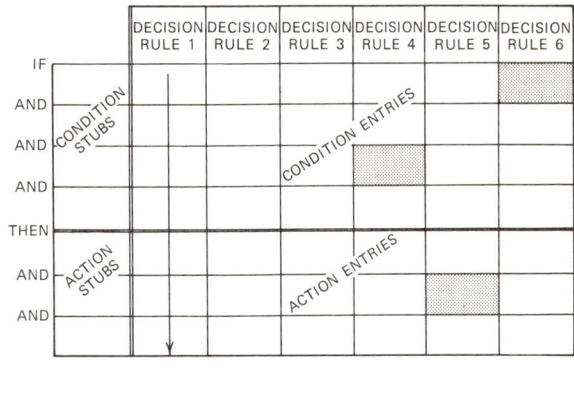

Figure 2-1 Decision table structure.

sents *if*, the remaining horizontal single lines represent *and*, and the horizontal double line represents *then*.

Limited Entry Rows

In the condition area (above the horizontal line) note the following:

"Y" prescribes that the condition in the stub must be satisfied;

"N" prescribes that the condition in the stub must not be satisfied;

"—" prescribes that it is immaterial whether the condition in the stub is satisfied or not.

In the action area (below the horizontal double line) note the following:

"X" prescribes the action in the stub that is to be executed if all the conditions of that rule are satisfied.

"—" prescribes that the action in the stub is to be ignored, whether or not all the conditions of that rule are satisfied.

The rules described in the decision table of Figure 2-2 are read as follows:

Decision Rule 1: If a customer's credit limit is satisfactory, approve his order.

Decision Rule 2: If the customer's credit limit is not satisfactory and his pay experience is favorable, approve his order.

Decision Rule 3: If the customer's credit limit is not satisfactory and his pay experience is not favorable and special clearance is obtained, approve his order.

Decision Rule 4: If the customer's credit limit is not satisfactory and his pay experience is not favorable and special clearance is not obtained, return his order to the sales department.

In this decision table (Figure 2-2) we have used only "Y," "N," "X," or "—" in the entry boxes. We call each of these rows a *limited entry* row.

	DECISION RULE 1	DECISION RULE 2	DECISION RULE 3	DECISION RULE 4
CREDIT LIMIT IS SATISFACTORY	Y	N	N	N
PAY EXPERIENCE IS FAVORABLE	–	Y	N	N
SPECIAL CLEARANCE IS OBTAINED	–	–	Y	N
APPROVE ORDER	X	X	X	–
RETURN ORDER TO SALES	–	–	–	X

Figure 2-2 Credit Policy.

Extended Entry Rows

An extended entry condition row contains part of the condition in the stub and the remainder in the entry portion of the rule. Extended-entry action rows are similar.

The extended entry style enables related conditions to be described in a more concise and easily stated manner; for instance, suppose the rules depend on various states (or values) of the variable X, such as in this limited entry:

STUBS...	RULES...
X = 1	Y N N
X = 2	- Y N
X = 3	- - Y

The same relationships can be described in extended entry form as

STUBS...	RULES...
X =	1 2 3

where the right-hand side of the condition stated in the stub is now placed directly in the rule entries in place of the "Y."

Extended entries may also be used for actions. This limited-entry style set of actions

STUBS...	RULES...
Add 10 to A	X - -
Add 10 to B	- X -
Add 10 to C	- - X

becomes, in extended entry, the following:

STUBS...	RULES...
Add 10 to	A B C

A decision table that contains limited-entry rows and extended-entry rows is known as a *mixed entry table*; Figure 2-3 is a sample mixed entry table.

	Rule 1	Rule 2	Rule 3	Rule 4	Rule 5	ELSE
IS EMPLOYEE NUMBER IN SEQUENCE	Y	Y	Y	N	-	-
IS DISCONTINUED DATE PRESENT	N	N	N	N	Y	-
EMPLOYMENT YEAR	NOT GREATER THAN (CURR-YR − 5 YRS)	LESS THAN CURR-YR	EQUAL TO CURR-YR	-	-	-
EMPLOYMENT YEAR	-	GREATER THAN (CURR-YR − 5 YRS)	-	-	-	-
VACATION WEEKS EARNED EQUAL TO	3	2	1	-	-	-
DO VACATION CALCULATION	X	X	X	-	-	-
WRITE ERROR	-	-	-	X	X	-
GO TO TABLE	PROCESS	PROCESS	PROCESS	READ	READ	ERROR

Figure 2-3 Sample mixed entry table (vacation week verification). The *Else*-rule in Figure 2-3 is the decision rule that is executed when the combined conditions of each of the other rules in this table are not satisfied.

Part II

THEORETICAL FOUNDATIONS

SECTION 3

Development of Decision Table Theory

INTRODUCTION

Decision tables can be used for such varied purposes as a tool for instruction, for the analysis and design of systems, for the solution of complex scientific problems, and for describing complex logic in computer programs. Whatever their purpose, decision tables require a sound theoretical basis to enable fuller exploitation of their potential. This chapter develops a particular theoretical structure.

As in all theory development, certain axioms and assumptions are the starting point. One of the key assumptions in the theory developed here is that one, and only one, of the rules in a decision table can be satisfied for each unique set of values for the variables of the conditions of that table. This does not preclude the development of a decision-table theory that assumes that any number of rules in the table can be satisfied by a given set of values for the conditions and in which the leftmost rule that is thus satisfied is then executed. The reader is cautioned, however, that theorems developed in one theoretical structure can not be carried over indiscriminately into another. With this brief introduction, we now develop the theory.

GENERAL

A decision table is based on a set of conditions, each of which may be evaluated as true or false at any given point in time. The truth or falsity of these conditions may be combined in various ways with each combination of conditions, along with its series of actions, known as a decision rule.

At any point in time the table may be evaluated in order to select that one decision rule corresponding to the particular combination of truth or falsity of the conditions at that point in time. The series of actions contained in the particular decision rule that has been selected is then executed.

Conditions

The symbol n represents the number of conditions (denoted by "C") present in the table. The subscript j is used to denote the sequence of conditions such that condition C_j may vary from 1 to n: $C_1, C_2, C_3, \ldots, C_n$.

Whenever evaluated, each condition C_j may be found to be either true or false, denoted by $a_j = 0$ (C_j is false) or $a_j = 1$ (C_j is true). This is indicated by saying that $V(C_j) = 0$ or $V(C_j) = 1$.

The status of the full set of n conditions is represented by a function:

$$S = (a_1, a_2, a_3, \ldots, a_n) \quad [3\text{-}1]$$

In the jth position of S, $a_j = 0$ [i.e., $V(C_j) = 0$] if C_j is false, and $a_j = 1$ [i.e., $V(C_j) = 1$] if C_j is true.

The Condition Structure

Each Condition, C_j, consists of two operands related by a relational operator. At least one operand is a condition variable[1], the other may be a condition variable or a constant. The relational operator may be any of the following: $=, \leq, \geq, <, >, \neq$.

At the time of evaluation of a truth value, $V(C_j)$ is obtained for C_j. Each a_j component of the S function is equal to $V(C_j)$; for example, consider the following four conditions:

$$C_1 = (W < 3),$$
$$C_2 = (X = 3),$$
$$C_3 = (Y > 4),$$
and $$C_4 = (Z \leq 0),$$

[1] A condition variable is a data item that can assume any one of a set of values.

where W, X, Y, and Z are condition variables. Representative values of these four variables are shown below together with the corresponding truth values of the associated conditions.

j	C_j	Actual Value of the Condition Variable	$V(C_j)$
1	(W < 3)	W = 2	1
2	(X = 3)	X = 5	0
3	(Y > 4)	Y = 3	0
4	(Z ≤ 0)	Z = −1	1

$S = (a_1, a_2, a_3, a_4) = (1, 0, 0, 1)$ in this case.

Condition Dependency and Independency

Dependence or independence exists between any pair of conditions. We define any two conditions, C_e and C_k, as being dependent if both have the same condition variable as an operand. Two conditions are independent if neither has the same condition variable as an operand.

There are two types of dependence: (a) mutual exclusion and (b) overlapping dependence.

MUTUAL EXCLUSION DEPENDENCY. Mutual exclusion dependency occurs for a pair of conditions C_k and C_e when there exists no value of the common condition variable such that both $V(C_k) = 1$ and $V(C_e) = 1$. Both $V(C_k)$ and $V(C_e)$ can never equal 1 at the same time. This is illustrated in Figure 3-1.

If X = 1 is true, 5 = X cannot be true at the same time; while if 5 = X is true, then X = 1 cannot be true. Note that both may be false at the same time, such as when X = 2.

C_k	C_e	Value of X	$V(C_k)$	$V(C_e)$
X = 1	5 = X	1	1	0
X = 1	5 = X	5	0	1
X = 1	5 = X	Not 5 and not 1	0	0

Figure 3-1 Mutual exclusion dependency.

We extend the definition of mutual exclusion to more than two conditions. Any number of conditions are mutually exclusive if at any point in time every two conditions in each of the pairs of conditions are mutually exclusive.

OVERLAPPING DEPENDENCY. Overlapping dependency occurs for a pair of conditions when there can exist at least one value of the common condition variable such that both C_k and C_e may be true. There may also be values that will cause other combinations of truth or falsity of the two conditions. This is illustrated in Figure 3-2.

DEPENDENCY DIAGRAMS. The possible relationships between $V(C_k)$ and $V(C_e)$ in the two types of dependency are illustrated in Figure 3-3.

It should be noted that such diagrams as are shown in Figure 3-3 may vary according to the nature of C_e and C_k; for example, in

C_k	C_e	Value of X	$V(C_k)$	$V(C_e)$
X > 0	X > 1	1	1	0
X > 0	X > 1	2	1	1
X > 0	X > 1	0	0	0

Figure 3-2 Overlapping dependency.

the case shown in Figure 3-2 in which $C_k = (X > 0)$ and $C_e = (X > 1)$, the combination $V(C_k) = 0$ and $V(C_e) = 1$ could never exist at the same time, that is, if $X > 1$, then X must be greater than 0.

Condition Requirements

Recall that two conditions are independent when neither has the same condition variable as an operand. Independent conditions that are required to be true or false in order for certain rules to be selected must be noted in the table as requiring specific

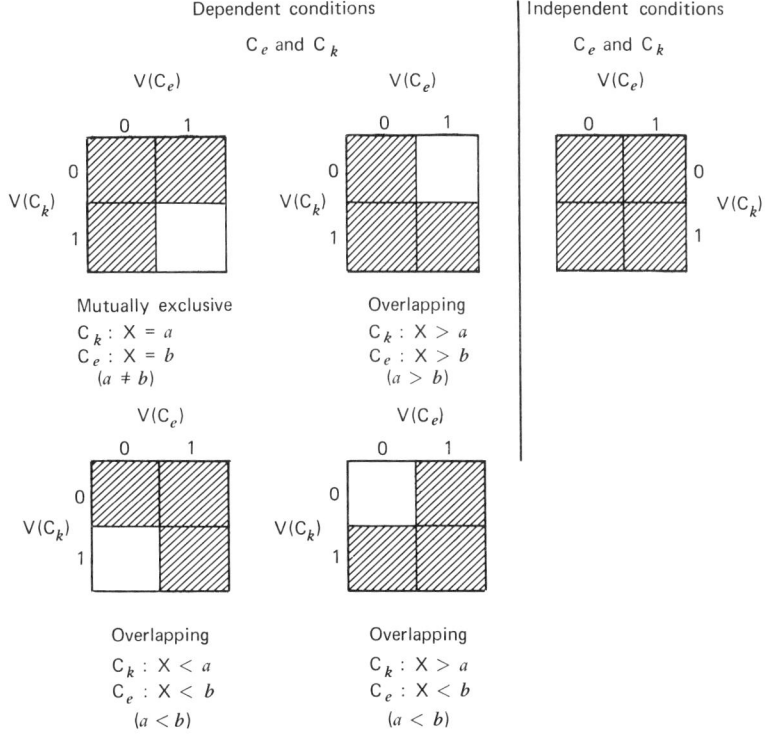

Figure 3-3 Examples of excluded combinations of true values in dependent and independent conditions.

tests upon the contents of their condition variable. The notation used is the following:

Y_i signifies that C_i *must* be proven true,
N_i signifies that C_i *must* be proven false.
$Y_i = \sim N_i$; $\sim Y_i = N_i$, where $\sim$ is the Boolean "NOT."

The Y_i and N_i are mutually exclusive and, together, include all possible states of the condition C_i. I_i denotes that a test on C_i is immaterial and C_i need not be proven either true or false.

$$I_i = Y_i + N_i$$
(where $+$ is an inclusive OR).

Recall that dependent conditions are a set of conditions that have at least one common condition variable. If the status of such a set of dependent conditions is immaterial for a specific rule, the I_i notation must be used in each condition of that set. However, if any one of the set has a value that is required to be either false or true, every member of the set should have one of the following condition requirements:

Y_i C_i is required to be proven true.   } explicit
N_i C_i is required to be proven false.

$*_i$ C_i is false, if some other explicit condition is proven.   } implicit
$\$_i$ C_i is true, if some other explicit condition is proven.

$*_i$ and $\$_i$ need not be proven false or true, respectively, for the condition to enter into the decision-making process. However, if tested, they have the full power of the N_i and Y_i respectively. Thus they are implicit but they are not immaterial.

AND-Functions

Let W_i be a variable representing Y_i, N_i, $*_i$, $\$_i$ or I_i for condition i. We now define an *AND-Function* B_j:

$$B_j = W_{1,j} \cdot W_{2,j} \cdot W_{3,j} \cdot \ldots \cdot W_{n-1,j} \cdot W_{n,j}$$
[3-2]
(where "$\cdot$" is the Boolean operator AND).

Each independent condition, C_i, may take on a requirement for W_i of Y_i, N_i, or I_i. Dependent conditions may also take on the implicit requirements $*_i$ and $\$_i$, but these may be viewed here as "special cases" of N_i and Y_i respectively.

Thus there are three states in which W_i may be expressed: Y_i, N_i, and I_i; and the number of possible forms of the AND-Function for n conditions is $j = 3^n$. A Table, T, which contains all $j = 3^n$ AND-Functions may now be defined:

$$T = \begin{bmatrix} B_1 \\ B_2 \\ \cdot \\ \cdot \\ \cdot \\ B_{3^n} \end{bmatrix}. \quad [3\text{-}3]$$

Each of the 3^n AND-Functions has a truth value of its own, denoted by $V(B_j) = \begin{Bmatrix} 1 \\ 0 \end{Bmatrix}$, depending on whether or not all n re-requirements of B_j are met.

Given a particular S function (see [3-1]), one can determine for that S whether $V(B_j) = 1$ or 0 by first substituting a 1 or 0 for each $W_{k,j}$ of B_j according to the rule indicated in Figure 3-4. If, for example, $W_{4,j} = N$ and $a_4 = 1$, then replace N in the 4th position of B_j by 0. If the resultant B_j contains all 1's, then $V(B_j) = 1$, otherwise $V(B_j) = 0$.

One more example is to have

$$B_j = (Y\ N\ *\ \$\ I\ N\ Y)$$
and
$$S = (1\ 0\ 0\ 1\ 0\ 0\ 1).$$

Then, replacing the Y, N, *, $, and I, according to Figure 3-4, we get

$$B_j = (1\ 1\ 1\ 1\ 1\ 1\ 1).$$

Hence $V(B_j) = 1$.

If $W_{k,j} =$	and $a_k =$	then replace $W_{k,j}$ in B_j by
Y	0	0
Y	1	1
N	0	1
N	1	0
*	0	1
*	1	1 (Note below)
$	0	1 (Note below)
$	1	1
I	0	1
I	1	1

Note: If a test is made on a condition with an implied $W_{i,j}$ but a conflicting a_k, then $W_{i,j}$ is treated as an explicit requirement, reversing the two cases noted above.

Figure 3-4 **Table of replacements for determining $V(B_j)$.**

As an illustration, consider the following set of conditions, in which the values of the condition variables are $X = 1$, $Y = 2$, $Z = 4$:

C_k	$V(C_k)$	
X = 1	1	
X = 3	0	and $S = (1, 0, 1, 0)$ [By 3-1]
Y = 2	1	
Z = 3	0	

The required values for these four conditions might be the following:

C_i	$W_{i,t}$
X = 1	Y
X = 3	*
Y = 2	I
Z = 3	N

Since $n = 4$, $j = 3^4 = 81$. There are 81 possible AND-Functions, but let us denote this one as the tth one.

$$B_t = Y_{1,t} \cdot *_{2,t} \cdot I_{3,t} \cdot N_{4,t} \quad [\text{By 3-2}]$$

Rewriting the S and B functions vertically to correspond to the format of Figure 3-4:

$B_{i,t}$	a_k	New $B_{i,t}$	(from Figure 3-4)
Y	1	1	
*	0	1	
I	1	1	
N	0	1	

Substituting the S function into the B function, using the rules stated in Figure 3-4, the following is obtained:

$$B_t = 1 \cdot 1 \cdot 1 \cdot 1$$
$V(B_t) = 1$, the function is true.

If the S function had contained any entry which, when combined with the corresponding $W_{i,j}$, produced a zero in the B_t AND-Function, then $V(B_t) = 0$ and the function would have been false.

Dependency of AND-Functions in a Table

Dependency of two AND-Functions is quite different from dependency of conditions. Two AND-Functions, B_k and B_t are dependent if for at least one set of values of the condition variables and condition requirements, both $V(B_k) = 1$ and $V(B_t) = 1$. Otherwise B_k and B_t are independent, that is, there exists no set of values of the condition variables such that both $V(B_k)$ and $V(B_t) = 1$.

EXAMPLE OF TWO DEPENDENT AND-FUNCTIONS. Suppose $B_5 = Y_1 \cdot N_2 \cdot Y_3 \cdot I_4$ (Note: $I_4 = Y_4 + N_4$), and $B_8 = Y_1 \cdot N_2 \cdot Y_3 \cdot N_4$. Then for that set of values of the condition variables that yields $S = (1\ 0\ 1\ 0)$, both $V(B_5) = 1$ and $V(B_8) = 1$. Then B_5 and B_8 are dependent.

EXAMPLE OF TWO INDEPENDENT AND-FUNCTIONS. Suppose $B_3 = Y_1 \cdot N_2 \cdot Y_3 \cdot Y_4 \cdot N_5$, and $B_7 = N_1 \cdot N_2 \cdot Y_3 \cdot Y_4 \cdot N_5$. The only set of values of the condition variables for which $V(B_3) = 1$, is the one that yields $S = (1\ 0\ 1\ 1\ 0)$. For this set of values, $V(B_7) = 0$. Therefore there exists no set of the condition variables such that both $V(B_3) = 1$ and $V(B_7) = 1$. Hence B_3 and B_7 are independent.

In general, then, if any i^{th} factor of a B_j function is a Y (or $) while the ith factor of a B_k function is an N (or *), the B_j and B_k are independent functions.

Inclusive OR-Functions

Using the function E_j to indicate the inclusive OR-Function that represents the relationship of the condition requirements U_i, we may write:

$$E_j = U_{1,j} + U_{2,j} + U_{3,j} + \ldots + U_{n-1,j} + U_{n,j}. \quad [3\text{-}4]$$

Hereafter, references to "OR" will mean "inclusive OR". Each condition, C_i, may take on a requirement, U_i, which may be Y_i, N_i or ϕ_i. Dependent conditions may also take on the implicit requirements of $*_i$ and $\$_i$.

Interpretation of each of these requirements is the same as those made for their use in the AND-Function, including the interpretation of $*_i$ and $\$_i$ as special cases of N_i and Y_i respectively. However, ϕ_i is the negation of I_i, that is, $\phi_i = \overline{I_i} = \overline{(Y_i + N_i)} = N_i \cdot Y_i$. Obviously, ϕ_i must be considered as a null requirement, since $N_i \cdot Y_i$ cannot occur.

The three states of U_i—N_i, Y_i, I_i—imply that a total of 3^n OR-Functions may be found for a set of n independent conditions. The table that represents all $j = 3^n$ OR-Functions is:

$$T' = \begin{bmatrix} E_1 \\ E_2 \\ \cdot \\ \cdot \\ \cdot \\ E_{3^n} \end{bmatrix} \quad [3\text{-}5]$$

Each of the 3^n OR-Functions has a truth value of its own denoted by

$$V(E_j) = \begin{Bmatrix} 1 \\ 0 \end{Bmatrix}.$$

$V(E_j) = 0$ if none of the requirements in E_j are met; $V(E_j) = 1$ if one or more requirements in E_j are met.

Transfer of $V(C_k)$ truth values from the S_j to the E_j function is similar to transfer from the S_j to the B_j function and is detailed in Figure 3-5.

If $U_{i,j} =$	and $a_k =$	then replace $U_{i,j}$ by
$\emptyset$	1	1
$\emptyset$	0	1
$\$$	1	1
$\$$	0	1 (Note below)
$*$	1	1 (Note below)
$*$	0	1
N	1	0
N	0	1
Y	1	1
Y	0	0

Note: If a test is made on a condition with an implied $U_{i,j}$ but a conflicting a_k, then $U_{i,j}$ is treated as an explicit requirement reversing the two cases noted above.

Figure 3-5 Table of replacements for determining $V(E_j)$.

To illustrate the inclusive OR-Function, E_j, as opposed to the AND-Function B_j, consider the following example:

While editing a certain field an error message is to be issued if at least one of these three tests fails: not numeric (C_1); has more than seven characters (C_2); has no decimal point (C_3).

Each of the conditions, C_i, may be true or false. The S function for the three conditions is

$$S = (a_1, a_2, a_3). \quad [\text{By 3-1}]$$

Three AND-Functions that would serve this purpose are

$$\begin{aligned} B_2 &= (Y_1 \cdot I_2 \cdot I_3), \\ B_3 &= (N_1 \cdot Y_2 \cdot I_3), \quad [\text{By 3-2}] \\ B_4 &= (N_1 \cdot N_2 \cdot Y_3). \end{aligned}$$

The OR-Function serving this purpose is

$$E_2 = (Y_1 + Y_2 + Y_3).$$

Relationships between AND and OR functions may be derived from De Morgan's theorem:

$$A \cdot B = \sim (\sim A + \sim B).$$

Derivations are not developed here.

Decision Table Notation

We now turn our attention to the notation used in decision tables. Tables that contain AND-Functions are the most popular and are discussed here; the discussion can be extended easily to tables that contain OR-Functions.

We assume that each table is based on a number, n, of conditions. Dependency in conditions is handled in the assignment of condition requirements, and in the more general case we assume conditions are independent. The actual number and kinds of conditions vary from table to table. We now consider the general case of a decision table based on n independent conditions, $C_1, C_2, C_3, \ldots, C_n$, each of which can be true or false. All notation previously described applies here.

Let D_j denote Decision Rule j, a_{ij} denote one action in D_j, and A_j denote the entire series of actions in D_j. Then

$$A_j = a_{1,j} \odot a_{2,j} \odot a_{3,j} \odot \ldots \odot a_{t,j};$$

again $j \leqslant 3^n$. The symbol "$\odot$" signifies that the actions $a_{1,j}, a_{2,j}, \ldots, a_{t,j}$ must be executed serially. $D_j = B_j \rightarrow A_j$ says, "Decision Rule j prescribes that if $V(B_j) = 1$, execute actions A_j."

A decision table is a structure for describing the expression

$$DT = D_1 \oplus D_2 \oplus \ldots \oplus D_{q-1} \oplus D_q; \quad q \leqq 3^n.$$

The symbol $\oplus$ signifies the Boolean "exclusive OR." One, and only one, D_j may be chosen. Hereafter the words "decision table" refer to both the structure and the expression, unless otherwise stated.

Definitions

A *pure* AND-Function is one that contains no I; for example, $P = Y \cdot * \cdot \$$ is a pure AND-Function.

A decision rule is *simple* if it contains a pure AND-Function; for example, if

$$P_1 = Y_{1,1} \cdot N_{2,1} \cdot N_{3,1} \cdot Y_{4,1} \cdot N_{5,1},$$

then $D_1 = P_1 \rightarrow A_1$ is a simple decision rule. A *mixed* AND-Function is one that contains one or more I's. A decision rule is *complex* if it contains a mixed AND-Function; for example, if

$$M_1 = I_{1,1} \cdot N_{2,1} \cdot Y_{3,1} \cdot N_{4,1} \cdot Y_{5,1},$$

then $D_1 = M_1 \rightarrow A_1$ is a complex decision rule.

Implicit Decision Rules

Decision rules can be implied in a decision table by one of two types of decision rules, either complex decision rules or ELSE-Decision-Rules. The ELSE-Decision-Rule (D_L) is defined as $D_L = ELSE \rightarrow A_L$.

If a decision table contains S simple rules and C complex rules, actions A_L are executed when a particular set of values of the condition variables yields a truth value 0 for each of the S simple rules and the C complex rules. This, of course, assumes that the totality of the S and C rules does not exhaust the set of possible rules for that decision table.

SECTION 4

Decision Table Theorems

INTRODUCTION

As a prelude to discussing the topics of decision table completeness, and contradiction or redundancy of decision rules, we present the supporting theorems with examples. The full development and proofs of the theorems are contained in Appendix I.

We remind the reader that the underlying axiom for the decision table theory developed here is that for any set of values for the variables in the conditions of a decision table, one, and only one, decision rule in that table can be satisfied by the particular set of values. The reader is further forewarned that relaxation of any part of the theory presented here automatically negates certain other portions of the theory.

RECAP OF DEFINITIONS

For convenience, we list previously defined and newly defined key terms that are used in the theorems that follow. A *pure AND-Function* is one that has no I's, that is, every one of its terms is either Y, N, *, or $.
A *mixed AND-Function* is one that is not pure; that is, it contains one or more I's.
A *simple decision rule* is one whose AND-Function is pure.
A *complex (or compound) decision rule* is one whose AND-Function is mixed.
ϕ is the OR-Function analogue of the I used in an AND-Function. ϕ is the requirement that the condition of the row it appears in does not have to be tested. Regardless of whether the value of that condition is TRUE or FALSE, the testing passes to other conditions.

Two functions (a pair of AND-Functions or a pair of OR-Functions) are *dependent* if for at least one set of values of the condition variables, the truth value for every pair of requirements is TRUE; for example, assume

AND-FUNCTION 1 = Y Y I

and

AND-FUNCTION 2 = I Y N.

For the set of values—
Y for Condition 1, Y for condition 2, and N for Condition 3—the truth value for Y and I (the first column above) is TRUE, the truth value for Y and Y (the second column above) is TRUE, and the truth value for I and N (the third column above) is TRUE.

Hence AND-Functions 1 and 2 are dependent.

THEOREMS FOR AND-FUNCTIONS AND OR-FUNCTIONS

In the theorems that follow, a Table, T, is assumed to comprise all AND-Functions and OR-Functions that can generate from the conditions of that table. The theorems designated by unprimed Roman Numerals refer to AND-Functions; those with single-primed Roman Numerals refer to OR-Functions; and those with double-primed Roman Numerals refer to both AND-Functions and OR-Functions. In the following examples we use AF to denote AND-Function, and OF to denote OR-Function. Also, a dash or blank is used for I.

THEOREM I. Within Table T, two AND-Functions are independent if, in at least one position, one function contains Y and the other function contains N. Otherwise, they are dependent.

THEOREM I′. Within Table T, two OR-Functions are dependent if, in at least one position, both functions contain a Y, or both functions contain an N. Otherwise, they are independent.

THEOREM I″. Within Table T, an AND-Function and an OR-Function are dependent if, in at least one position, both functions contain a Y, or both functions contain an N, or the AND-Function contains an I, and the OR-Function contains a Y or N. Otherwise, they are independent.

For illustration of the above three theorems, consider the following table.

	AF1	AF2	AF3	OF1
CUST-CODE = 'SPECIAL'	Y	–	N	N
ACCT-RECV-AMT > 5000	–	N	Y	ϕ
ACCT-RECV-AGE > 120	–	–	N	ϕ

AF1 and AF2 are dependent—there does not exist at least one Y,N pair for the three conditions.

AF1 and AF3 are independent—a Y,N pair exists for the first condition.

AF2 and AF3 are independent—a N,Y pair exists for the second condition.

AF1 and OF1 are independent—a Y,Y or a N,N pair does not exist, nor does there exist both an I for the AF1 and a Y or N for the OF2 for each of the three conditions.

AF2 and OF1 are dependent—an I (or dash) exists in the first condition of AF2 and an N exists in that first condition for OF1.

AF3 and OF1 are dependent—an N,N pair exists for the first condition.

THEOREM II. Within Table T, each pure AND-Function is independent of every other pure AND-Function.

Corollary of Theorem II. Within Table T, there exist exactly 2^n pure AND-Functions. Each of the remaining $[2(3^n) - 2^n]$ functions is either a mixed AND-Function or an OR-Function.

For illustration of Theorem II and its corollary, consider the following table comprising two conditions.

	AND-FUNCTIONS									OR-FUNCTIONS								
	1	2	3	4	5	6	7	8	9	1	2	3	4	5	6	7	8	9
Salaried employee?	Y	Y	N	N	Y	–	N	–	–	Y	Y	N	N	Y	φ	N	φ	φ
Worked overtime?	Y	N	Y	N	–	Y	–	N	–	Y	N	Y	N	φ	Y	φ	N	φ

The first four AND-Functions are the pure AND-Functions of this table. They are all independent of each other according to Theorem I. The remaining $[2(3^n) - 2^n] = [2(3^2) - 2^2] = 14$ functions are mixed AND Functions or OR-Functions. One cannot conceive of any other AND-Function or OR-Function that doesn't duplicate one of the functions already in the above table.

THEOREM III. The form of a mixed AND-Function that contains I in r positions ($1 \leq r < n$) can be expanded into a canonical form that consists of 2^r pure AND-Functions each connected by an exclusive "OR" operator.

Corollary 1 of Theorem III. The canonical form of a mixed AND-Function contains an even number of pure AND-Functions.

Corollary 2 of Theorem III. The canonical form of a mixed AND-Function contains at least two pure AND-Functions.

THEOREM III'. An OR-Function that contains φ in r positions ($0 \leq r < n$) can be converted to (is equivalent to) $(2^n - 2^r)$ distinct pure AND-Functions.

Corollaries 1 and 2 or Theorem III are evident from Theorem I that specifies 2^r, with $r \geq 1$. For illustration of Theorems III and III', consider the following table.

AF1 has 1 dash (equivalent to I) and can be expanded to $2^1 = 2$ pure AND-Functions:

AF1a	AF1b
Y	Y
Y	Y
Y	N

OF1 has 2 φ's and can be expanded to $(2^3 - 2^2) = 4$ distinct pure AND-Functions:

AF4	AF5	AF6	AF7
N	N	N	N
Y	Y	N	N
Y	N	Y	N

THEOREM IV. Within Table T, every mixed AND-Function that contains I in r positions ($1 \leq r < n$) is dependent on each of 2^r pure AND-Functions of T.

Corollary 1 of Theorem IV. Two mixed AND-Functions are dependent on each other if their canonical forms each contain one or more pure AND-Functions that are common to both.

Corollary 2 of Theorem IV. A mixed AND-Function is dependent on each of the pure AND-Functions contained in its canonical form.

Corollary 3 of Theorem IV. If a pure AND-Function and a mixed AND-Function are dependent, the pure AND-Function is contained in the canonical form of the mixed AND-Function.

	AF1	AF2	AF3	OF1
Credit approved?	Y	Y	Y	N
Inventory available?	Y	N	N	φ
Backorders allowed?	–	Y	N	φ

Corollary 4 of Theorem IV. If two mixed AND-Functions are dependent, there exists at least one pure AND-Function in their canonical form that is common to both.

THEOREM IV‴. Within Table T, every OR-Function that contains ϕ in r positions ($0 \leq r < n$) is dependent on $(2^n - 2^r)$ distinct pure AND-Functions.

Consider the following table.

	AF1	OF1
Field has more than four positions?	N	Y
Field is not numeric?	Y	ϕ
Field has decimal point?	–	ϕ

Table T would contain AF1 and the following pure AND-Functions to which it expands.

AF1a	AF1b
N	N
Y	Y
Y	N

Referring back to Theorem I, AF1 is dependent on AF1a and AF1b. Thus Theorem IV is illustrated.

OF1 can be expanded to the following AND-Functions:

AF2	AF3	AF4	AF5
Y	Y	Y	Y
Y	Y	N	N
Y	N	Y	N

Referring back to Theorem I‴, OF1 is dependent on AF2 through AF5. Thus Theorem IV‴ is illustrated.

To illustrate Corollary 3 of Theorem IV, we use the following example.

	AF1	AF2
Premium overdue?	Y	Y
Special authorization?	N	–

Referring to Theorem I, AF1 and AF2 are dependent. We expand AF2 to the following:

AF2a	AF2b
Y	Y
Y	N

As prescribed in Corollary 3, AF1 is contained in AF2, namely AF2b. Similar illustrations can be developed for Corollaries 1, 2, and 4 of Theorem IV, but are left as an exercise for the reader.

THEOREM V. Table T, based on n conditions, contains one, and only one, set of 2^n independent pure AND-Functions.

As illustration of Theorem V, consider the following table:

	AF1	AF2	AF3	AF4	AF5	AF6	AF7	AF8
American citizen?	Y	Y	Y	Y	N	N	N	N
Reached voting age?	Y	Y	N	N	Y	Y	N	N
Convicted of felony?	Y	N	Y	N	Y	N	Y	N

With the number of conditions equal to 3, the total number of pure AND-Functions should be $2^3 = 8$, according to Theorem V, and is illustrated above. No other pure AND-Function exists that does not duplicate one of those in the table.

THEOREM VI. A complex decision rule (a rule that contains mixed AND-Functions) that contains 1 in r positions ($1 \leq r < n$) of its AND-Function is equivalent to 2^r simple decision rules (rules that contain pure AND-Functions).

THEOREM VI′. A decision rule that contains an OR-Function with ϕ in r positions ($0 \leq r < n$) is equivalent to $(2^n - 2^r)$ simple decision rules.

The above theorems are an extension of Theorems III and III′. One need only append actions to each of the AND-Functions and OR-Functions of the tables used to illustrate Theorems III and III′ to form a decision table that illustrates Theorems VI and VI′.

PART III

ANALYSIS AND IMPLICATIONS OF DECISION TABLES

SECTION 5

Conditions: Criteria For Decisions

The relationship of conditions in a decision table rule can be viewed as a conjunction of the individual conditions (conditions connected by AND) for that rule;[1] for example, IF NOT cond-1 AND cond-2 AND NOT cond-4 AND cond-6, THEN execute a series of actions. The omitted conditions (cond-3 and cond-5) were "don't-cares." Those containing the qualifier "not" were N entries; the others were Y entries.

Consider now an example of what these conditions might be, as shown in Figure 5-1.

STUBS...	RULE ENTRIES...	
X = 4	Y	N
X = 6	–	Y
U = 1	Y	–
Z = 0	Y	N
Q = 1	N	–
Z = 2	–	Y

Figure 5-1 Sample condition section.

The first rule might be stated as:
IF X = 4 AND U = 1 AND Z = 0 AND Q NOT = 1, THEN ...
The second rule might be stated as:
IF X NOT = 4 AND X = 6 AND Z NOT = 0 AND Z = 2,
 THEN ..., where the actions for the proper rule (not shown here) follow the "THEN".[2]

[1] Some decision table methods permit the relationships between conditions to be other than AND; namely, OR.

[2] Decision table preprocessors do not generate such long IF statements, of course. Decomposition, the process that generates IF statements, is discussed elsewhere.

Although the IF statement stated above is not logically wrong, it does contain redundant checks on the values of X and Z. That is, if X must equal 6, it cannot possibly be equal to 4 at the same time. The same sort of argument holds for the two relations of Z. Such inefficiency is intolerable.[3]

Traditionally, decision tables have not allowed directly for the elimination of such redundant checks, although decomposition techniques minimize their effect.

BASIC ASSUMPTIONS OF THE TRADITIONAL DECISION TABLE CONDITION

The above situation is a result of the two basic assumptions historically made about AND-connected conditions of decision tables. The first assumption has been that conditon rows are unrelated to one another.

The resolution of one condition for a decision rule tells us nothing about the possible resolution of any other condition for that rule.

This assumption is basic to the highly important task of optimization.

The second assumption evolves from the postulate that the passing against the table of any one transaction must result in the selection of one and only one rule and that the ELSE-rule is available to assure completeness of the set of rules. This has historically given rise to the general acceptance by decision table users that the resolution of any one condition must result in a con-

[3] Decomposition techniques usually are designed to avoid the more obvious inefficiencies, but some cannot be avoided without specific allowances for dependencies.

tribution to narrowing down the set of eligible rules that is independent of the contribution of each of the other conditions. Of course, the testing of all conditions taken together must narrow down to a single rule.

It is this contribution of a single condition to the process of rule selection that is reflected in traditional decision tables as the second of the assumptions.

For a single limited entry condition row, each entry must be either identical to, or mutually exclusive with, any other entry for that condition, or it must be a don't-care.

This mutual exclusion is usually represented by allowing only the rule entries Y, N, and blank (or often a dash). Y and N are mutually exclusive (a binary Yes or No resolution of the condition) and the blank represents either Y or N, or simply don't-care. Extended entries are easily restated in terms of limited entry, as illustrated in Figures 5-3 and 5-4.

As a consequence of the second assumption, the resolution of a certain condition (i.e., we evaluate the expression during a transaction) for any one non-don't-care entry automatically resolves it for all entries. Consider, for example, the condition row in Figure 5-2.
Once the status code in a given transaction is tested for the first rule (Y) and found to be either greater or not greater than zero, then every other rules' entry for this condition is known without further testing of the expression in the stub.

Traditional Limited Entry

We denote as *Traditional Limited Entry* those condition rows made up of conditions

STUB....	RULE ENTRIES....
STATUS-CODE GREATER THAN ZERO	Y N N N Y – – N

Figure 5-2 Sample of complete determination of rule entries by one test of the condition.

which may be resolved in a binary Yes-No manner only once to determine the effect of the rule entries; these rule entries are Y, N, or "don't cares." "Traditional" refers to the historical popularity of conditions which utilized the mutually exclusive nature of the two non-don't-care (Y, N) rule entries. "Limited Entry" refers to the fact that the entire statement of the condition is in the stub with the rule entries containing only indicators (Y, N, —) to describe the state that each condition in a rule must be in before moving to the next condition in the rule.

STUBS...	RULE ENTRIES...				
VERB = 'ADD'	Y	N	N	N	N
VERB = 'SUBTRACT'	—	Y	N	N	N
VERB = 'MULTIPLY'	—	—	Y	N	N
VERB = 'DIVIDE'	—	—	—	Y	N
VERB = 'COMPUTE'	—	—	—	—	Y

Figure 5-4 Limited entry condition rows converted from extended entry.

Traditional Extended Entry

Condition rows need not be just limited entry; that is, the condition might start in the stub and then finish in the rule entry itself. This type of condition row is known as *Extended Entry*. Typically, such an entry has several different "endings" in the entries as in Figure 5-3.

An extended-entry condition row does not have just one expression, as does the limited-entry condition row. The expression of the condition is generally different for each rule entry. Each entry is either mutually exclusive with, or identical to, each of the other entries, or they are don't-cares. Checking one rule's condition does not, in general, allow a determination of whether the condition of each of the other rules is true or false. However, the above extended entry can be easily rewritten as five traditional limited-entry condition rows, as shown in Figure 5-4.

Figure 5-4 uses the traditional technique for writing limited entry rows to enable decomposition of the table by certain preprocessors. Again, popular historical usage of this type of entry suggested the use of the name "Traditional."

Characteristics of Traditional Conditions

Both traditional types of condition rows —traditional limited and traditional extended entry—share the property that each entry is either mutually exclusive or identical to any of the other entries, or is a don't-care. The other property shared by traditional condition rows is that they are considered unrelated to ("independent" of) any of the other condition rows in the table.

Related Conditions

In reality condition rows may be related to one another. We have already seen examples in which the occurrence of mutually exclusive checks on the same variable were made simply because the presence of the same variable in several conditions is not provided for in the traditional condition concept. In such cases, the recognition of relatedness of condition rows (or as more commonly referred to, dependencies of conditions) is only an aid to efficiency.

STUB...	RULE ENTRIES...				
VERB =	'ADD'	'SUBTRACT'	'MULTIPLY'	'DIVIDE'	'COMPUTE'

Figure 5-3 Sample extended entry condition row.

30 CONDITIONS: CRITERIA FOR DECISIONS

Mutually Exclusive Conditions

The preceding example (see Figures 5-3 and 5-4), in which five possible cases of the data-item VERB were specified, represents a limited entry group of conditions that are so related that only one of the conditions can be satisfied at any one time (i.e., the conditions are mutually exclusive). If VERB equals "ADD," it obviously does not equal "COMPUTE" at the same time. Most preprocessors will handle such a simple structure efficiently. However, if it is embedded within a more complicated structure, redundant comparisons will sometimes occur because of the presence of both Y's and N's in a single rule.

We can replace the N's by don't-cares (dashes), to yield Figure 5-5. However, this will not guarantee the elimination of redundant checks on VERB, inasmuch as these five conditions are assumed to be unrelated. Thus a true finding of any one of the five checks does not help narrow down the rule to be finally selected, as was possible with the N's in the original limited-entry example. This role of the N is discussed later in the section *Rule: Redundancy and Contradiction.*

The Asterisk and Dollar Sign Entries

The Asterisk (*) and Dollar Sign ($) entries are special rule entries to indicate mutual exclusion of one condition with another on a rule-by-rule basis. The asterisk entry appears in the rule entry position of a limited entry condition to indicate that that condition is required to be *false* for that rule and that some other condition for that same rule is adequate to satisfy the requirement. The $ indicates that a condition is required to be *true,* with some other condition available to insure satisfaction of that requirement.

In other words, an asterisk acts as an N, but is implicit rather than explicit. Likewise, the dollar sign acts as a Y, but is implicit. They require no explicit check to ensure the condition, but if a check is made for other purposes, the * has the power of an N and the $ has the power of a Y.

In the use of * or $ (unlike the N or Y entry), no advantage ensues from using the minimum number of *'s or $'s required to be definitive, while leaving the remainders to don't-cares. Indeed, indicating all of the mutually exclusive situations with asterisks or dollars is advantageous both in theory and practice. Using our previous example, the asterisk form would be as shown in Figure 5-6.

To illustrate the asterisk's role, the Y in row 3 of the third rule indicates that the value 'MULTIPLY' must be present for that rule to be selected. The asterisks in the remaining rows of Rule 3 indicate that the other four conditions are impossible if the Y for condition 3 is satisfied. The implication here is that no special checks need be made for those other four values. Whenever one of the related conditions is tested and found to be true (satisfying a Y rule entry), all asterisk entries on the same con-

STUBS...	RULE ENTRIES...				
VERB = 'ADD'	Y	–	–	–	–
VERB = 'SUBTRACT'	–	Y	–	–	–
VERB = 'MULTIPLY'	–	–	Y	–	–
VERB = 'DIVIDE'	–	–	–	Y	–
VERB = 'COMPUTE'	–	–	–	–	Y

Figure 5-5 Table of rules with N replaced by dash (an incorrect table).

STUBS...	RULE ENTRIES...				
VERB = 'ADD'	Y	*	*	*	*
VERB = 'SUBTRACT'	*	Y	*	*	*
VERB = 'MULTIPLY'	*	*	Y	*	*
VERB = 'DIVIDE'	*	*	*	Y	*
VERB = 'COMPUTE'	*	*	*	*	Y

Figure 5-6 A table whose rule entries contain *.

dition row imply that these rules that contain asterisks *cannot* be selected. This latter effect, that of the horizontal effect of the asterisks, is best understood in terms of decision table decomposition techniques. The use of the dollar sign is illustrated in a later discussion of bounded ranges.

Inequalities in Conditions

Inequalities in conditions represent cases in which a variable assumes a value within a range of values, rather than an explicit value. Inequalities include *greater, less, not equal,* as well as *between a and b* type of cases. Any inequality can be expressed as the latter type (between *a* and *b*), with *a* and *b* denoted as *bounds* and all of the values satisfying that inequality as a *range*. In practice, however, it is generally convenient and practical to use the other forms of inequality; we take that approach here.

Using Ranges of Values as Conditions: Bounded Ranges

We now consider another type of relationship between conditions—bounded ranges.

A variable may be interpreted, insofar as its rule selection function goes, according to several ranges of values, rather than on a specific value; for example, in a quality-control problem we may wish to take different actions depending on the ratios of two materials. The use of bounded ranges to de-

	STUBS...	RULE ENTRIES...					
C	AGE GT	0	5	15	18	16	20	21
C	AGE LE	5	18	16	21	20	21	–
C	SEX =	–	F	M	F	M	M	–
A	RATING =	0	1	1	2	2	3	3

Note: Again GT and LE are special operators denoting GREATER THAN and LESS THAN or EQUAL TO, respectively.

Figure 5-8 A decision table with overlapping ranges.

scribe this type of problem is illustrated in Figure 5-7.[4]

Nonoverlapping and Overlapping Ranges

Notice in Figure 5-7 that no pair of ranges for each of the rules overlaps; thus these ranges are mutually exclusive with one another. This case of nonoverlapping ranges is therefore similar in application to traditional extended entry, if we consider the two bounded-range condition rows as one row.

Overlapping ranges are different in both theory and practice. The use of overlapping ranges in a decision table is illustrated in Figure 5-8. Ratings are assigned to individuals on the basis of sex and age. The individual's characteristics (age and sex) are entered as transactions that are passed

[4] The reason for using the non-COBOL operators GT (Greater Than) and LE (Less Than or Equal) will be made clear shortly. These operators are called bound-operators, which also include LT, GE, and NE.

	STUBS...	RULE ENTRIES...						
C	RATIO GT	0	6	7	7.4	7.5	8.0	9.4
C	RATIO LE	6	7	7.4	7.5	8.0	9.4	100
A	QUALITY =	"P"	"F"	"G"	"E"	"G"	"F"	"P"

Note: GT and LE are special operators denoting GREATER THAN and LESS THAN or EQUAL TO, respectively. C on the left indicates condition; A indicates action.

Figure 5-7 Table with ranges of values for conditions.

against the rules of the Figure 5-8 decision table to determine his rating.

Notice that the age ranges for each rule are not mutually exclusive with each other, in apparent violation of the two basic assumptions of traditional decision tables. However, the total table still maintains its logical integrity, since every rule is independent of the others; that is, given a particular age and sex, one and only one rule in the table will be satisfied.

The don't-care (dash) in the second row indicates that AGE is to be less than or equal to the whatever upper limit age can be attained. The use of this dash in practice is much more obvious than any theoretical discussion might portray.

Neither of the first two rows of Figure 5-8 are traditional as they stand—a check with a true or a false result of any one of the entries in either row does not provide enough information to narrow down the selection of the proper rule. Assume, for example, a given transaction provides a Yes answer to the condition of the first rule, AGE greater than 0. This does little to determine the rule to be selected and thus cannot be considered as meeting the Yes, No, or don't-care criteria for limited entry or the mutually exclusive conditions of extended entry.

Handling the above bounded conditions requires special recognition and processing on the part of the decision table preprocessor. Special recognition can be based on the use of such special non-COBOL operators GT, GE, LT, LE, and NE (Greater Than, Greater or Equal, Less Than, Less Than or Equal, and Not Equal, respectively); we denote these five operators as *bound-operators*. Usually, a pair of bound-operators that includes the equal case on one side but not on the other is used, for example, GT and LE.

STUBS...	RULE ENTRIES...				
X NOT =	1	2	3	4	5
X NE	1	2	3	4	5

Note: These two condition rows are not equivalent.

Figure 5-9a   Two types of nonequal conditions.

APPROACHES TO THE "NOT EQUAL" CASE. A special bound-operator, NE, might be designed to eliminate individual specific values from within ranges. It may also be used to eliminate specific values of a variable from nonrange situations.

It will be instructive at this point to compare the bound-operator NE to the use of a COBOL operator "NOT =" in a nonbounded extended entry. Each of these two condition rows is shown in Figure 5-9a.

The use of "NOT =" in the first condition of Figure 5-9a illustrates an improper usage. For the first row, NOT = will be evaluated as a traditional extended entry, in which each rule's condition is mutually exclusive from every other rule's condition unless their entries are identical or don't-cares. Thus in the "NOT =" case, if X is first evaluated for "X NOT = 1" and found true (e.g., X = 4), then Rule 1 will be considered true, the others false; but if "NOT = 5" were to have been evaluated first, the fifth rule would have been found true and the first four would have been considered false. Obviously, rule selection would be a matter of chance rather than the orderly procedure it must be.

In the second row of Figure 5-9a which uses the NE (a bound-operator), if X NE 1 is checked first and found to be true (again, use X = 4 as a sample case), the other rules would *not* be considered automatically false as a result. This is because the conditions associated with the bound-operator NE are not considered "either mutually exclusive or identical" as are traditional extended conditions. The NE bound-operator requires

BASIC ASSUMPTIONS OF TRADITIONAL DECISION TABLE CONDITION

STUBS...	RULE ENTRIES...				
X GT	1	7	12	–	1
X LE	7	12	16	–	7
X NE	3	–	–	2	–
Z = 4	Y	Y	Y	N	Y
Y = 2	Y	–	–	–	N

Note: This figure illustrates exclusion of the specific value of X = 3 from the range 1-7 for the first rule but not for the fifth rule. The value of X = 2 is eliminated from the fourth rule. The Z and Y conditions ensure that the rules will be independent.

Figure 5-9b Illustration of the bound-operator NE.

only one row, rather than the two rows required by other bound-operators.

Use of Don't-Care Entries in Bounded Ranges

Don't-care entries are permitted in bounded condition rows but where they appear for a rule they must be matched on each row involved in the definition of a single range. The bound-operator NE is an exception; it may exclude specific values by having blank entries in addition to those indicated in the GT, LT, GE, or LE condition rows defining the range (as illustrated in Figure 5-9b).

Don't-care (dash) entries may also be used to represent the limit of a singly-bound range, as was illustrated in the second row of the last rule of Figure 5-8.

Overlapping Conditions with Other than Bound-Operators

Traditional conditions specify situations that are either identical or mutually exclusive between rules. As shown in Figure 5-8, many cases exist in which we might wish to specify situations in which overlap occurs. In those conditions—considered as ranges—neither identity nor mutual exclusion exists between rules. Obviously such overlapping conditions cannot be used to differentiate between rules for the purpose of rule selection but rather other conditions must carry this burden.

A case of such a usage is the specification of limits where the limits are peculiar to the rule and represented by singly bounded inequalities. An example is seen in Figure 5-10, in which traditional entry conditions were improperly used to express the situation. The special operator, LT, should have been used.

In this case reorder procedures will be executed whenever inventory of a part falls below the prescribed levels for the various part-types. The set of part-types provides the rule selection criteria and is expressed in a traditional extended condition row. However, the INVENTORY item is to be evaluated for values that are not mutually exclusive between rules, while the condition row coding for inventory assumes such mutual exclusion and is incorrect.

The second bound on the inventory range is assumed to be not less than zero for all rules.

A person reading the table can arrive at the correct decision rules. However, a preprocessor cannot, unless it receives an indicator of some sort that says when INVENTORY is less than 275, check Part Type C. If not C, then try to find Rule 1 by checking

	STUBS...	RULE ENTRIES...			
C	INVENTORY LESS THAN	15	4	275	12
C	PART-TYPE EQUAL TO	"A"	"B"	"C"	"D"
A	PERFORM RE-ORDER	X	X	X	X

Figure 5-10 Table containing overlapping conditions.

34 CONDITIONS: CRITERIA FOR DECISIONS

whether INVENTORY is less than 15, and so forth.

Range Handling Using Traditional Entries Only.

This example (in Figure 5-10) obviously needs the use of bound-operators to be properly handled. But since current operational decision table preprocessors do not offer full bound-operator capabilities for overlapping ranges, an approach using traditional operators is highly desirable. The following discussion presents such an approach, using the * and $ operators. If a particular preprocessor does not yet support these two operators, the N may be used for the *, while the Y may be used for the $ with effects only in the area of efficiency of the computer code generated by the preprocessor.

The use of $ in conjunction with limited entry conditions is best illustrated with two examples: one for a Less Than situation, the second for a Greater Than situation. Note that the special non-COBOL operators mentioned in connection with bounded entry condition rows must not be used here; only the regular COBOL operators can be used. For the case just discussed for INVENTORY (Figure 5-10), write the limited entry condition rows with the greatest value first, placing a Y in the rule entry corresponding to the rule(s) which that value affects (as shown in Figure 5-11).

STUBS...	RULE ENTRIES...			
INVENTORY @ 4	–	Y	–	–
INVENTORY @ 12	Y	–	–	Y
INVENTORY @ 275	–	–	Y	–
PART-TYPE =	'A'	'B'	'C'	'D'

Note: If the operators were LESS THAN, *the @ refers to* LESS THAN; *if the operators were* GREATER THAN, *the @ refers to* GREATER THAN.

Figure 5-11 First step in using the $ entry.

STUBS...		RULE ENTRIES...			
C	INVENTORY < 4	–	Y	–	–
C	INVENTORY < 12	Y	$	–	Y
C	INVENTORY < 275	$	$	Y	$
C	PART-TYPE =	'A'	'B'	'C'	'D'
A	PERFORM RE-ORDER	X	X	X	X

Figure 5-12 Second step in using the $ entry for less than.

$ ENTRY: OVERLAPPING LESS THAN CASE. In the conversion of Figure 5-11 using LESS THAN (denoted by the symbol <), the $ is filled in below the Y in each rule column (only for those condition rows that are related). This produces the table shown in Figure 5-12, which safely represents, without the use of bound-operators, the logic intended in Figure 5-10.

To further illustrate the role of the $ entry, consider the first rule in Figure 5-12. If the inventory of part-type A is less than 12, it is obviously also less than 275 (indicated by the $). It is immaterial whether or not it is less or greater than 4 (don't-care dash).

$ ENTRY: OVERLAPPING GREATER THAN CASE. In the conversion of Figure 5-11, using GREATER THAN (denoted by the symbol >), the $ is filled in above the Y in each rule column (only for those condition rows that are related). This is shown in Figure 5-13.

STUBS...	RULE ENTRIES...			
INVENTORY > 4	$	Y	$	$
INVENTORY > 12	Y	–	$	Y
INVENTORY > 275	–	–	Y	–
PART-TYPE =	'A'	'B'	'C'	'D'

Figure 5-13 Second step in using the $ entry for greater than.

FULL OVERLAPPING RANGES. With respect to the $ entry, we have thus far discussed ranges where one bound is implied. General ranges in which two bounds may be stated to define ranges without regard to how ranges for one rule correspond with those for any other rule, may be handled by properly and efficiently using both the $ and * entry. Discussion of the coding techniques is somewhat longer and is deferred until the section on *Writing the Decision Table*.

Relation of Data Item to Data Item

So far all of our operations involving inequalities discuss data items compared with literals. We now complete the discussion of the use of inequalities by including the relation of a data item to another data item.

First, consider the rather classical case of comparing one data item to another with a test on Greater Than, Equal To, or Less Than to achieve one of three mutually exclusive results. This may be handled by either (a) limited entry conditions, with or without the asterisk entry, or (b) extended entry conditions in which the operator is included in the rule entry along with the right-hand operand, or (c) by use of the special substitution operator ("colon operator") a description of which follows.

STUB...	RULE ENTRIES...
ITEM-A : ITEM-B	< > =

Figure 5-14 Sample of condition row using the colon operator.

Colon Operator for Comparison of Data Item to Data Item

The colon operator appears between the two data items in the stub. One of the three symbols <, >, or = appears in each rule entry. The entry for a particular rule replaces the colon in the stub, and that condition expression is the condition for that rule.

This comparison of two data items to achieve one of the three mutually exclusive results fits into the traditional condition row situation quite well, as shown in Figure 5-14.

This is the case in which one data-item is being compared with another data-item in an extended condition row using whatever inequalities the problem requires.

The use of the colon operator can be useful in a payroll tax calculation as shown in Figure 5-15. If Y-TO-D-FICA is greater than the maximum (LIMIT), enter a special routine to reduce this period's deduction (1st rule); if less, calculate FICA (2nd

	STUBS...	RULE ENTRIES...		
C	Y-TO-D-FICA : LIMIT	>	<	=
A	GO TO	REDUCE-FICA	CALC-FICA	BYPASS-FICA

The above table may be written without the colon entry as

C	Y-TO-D-FICA > LIMIT	Y	*	*
C	Y-TO-D-FICA < LIMIT	*	Y	*
C	Y-TO-D-FICA = LIMIT	*	*	Y
A	GO TO	REDUCE-FICA	CALC-FICA	BYPASS-FICA

Figure 5-15 An example of the use of the colon operator and an equivalent limited entry table.

rule); if equal, bypass further processing (3rd rule).

Inequalities Between Two Data Items

The use of the NOT EQUAL between two data items is relatively straightforward; it fits quite well into the traditional limited entry usage. A "NOT EQUAL" between two data items can always be converted to an "EQUAL" case by reversing the Y and N limited rule entries, as well as the *, $ entries:

A NOT EQUAL B   Y – N N Y $ *
is equivalent to
A EQUAL B   N – Y Y N * $

Greater or Less Than relations between two data items also fit into traditional limited entry usage. The situation here is quite similar to that encountered with the colon operator. Figure 5-15 illustrates a two-data-item inequality situation. We now consider multiple data items.

Inequalities Between Multiple Data Items

Let us first consider the case in which only one of the extended entries can meet the condition at a time (that one may, of course, be duplicated by identical entries any number of times). An example (Figure 5-16) is the checking for a nonblank field among a number of fields when, for some external reason, it is known that only one of those fields will be nonblank.

In this case the situation reduces to the equivalent of the traditional limited entries shown in Figure 5-17.

STUB...	RULE ENTRIES...
SPACE LESS THAN	DN1 DN2 DN3 DN4

Figure 5-16 Condition row with data items in entries.

STUBS...	RULE ENTRIES...			
SPACE LESS THAN DN1	Y	*	*	*
SPACE LESS THAN DN2	*	Y	*	*
SPACE LESS THAN DN3	*	*	Y	*
SPACE LESS THAN DN4	*	*	*	Y

Figure 5-17 Condition row (of Figure 5-16) converted to limited entry.

The second, and more important, case of the use of the Greater and Less inequalities against several data items is when the individual comparisons may or may not all be true. This case is similar to the situation we met when discussing the NOT EQUAL case earlier in this section.

The use of regular extended entry, with its assumption that all entries are either identical or mutually exclusive, is ruled out and no special provision can be made for this case.[5] The best way to handle this situation is to use limited entry condition rows for each, indicating Y for those rules involved, with don't-cares (dashes) for all other rules as shown in Figure 5-18.

This type of structure eliminates the requirement that the inequality conditions must make each of the rules unique; other conditions can carry this load. In the example below, the last condition—a traditional extended entry—insures that the table that incorporates overlapping ranges has independent rules.

The following condition-related features are either implemented in available preprocessors, or are not implemented but desirable:

1. Local ELSE clauses that permit a single table to be subdivided with residual rules for various groups of conditions. The

[5] The GT, LT, GE, LE, and NE bound-operators do not provide a technique for handling the Greater or Less relationships between data-names. It is conceivable that these special bound-operators could be redefined to work with bounds which can be varied during execution.

Not this

STUBS...	RULE ENTRIES...			
DN1 GREATER THAN	DN2	DN3	DN4	DN5
Q =	'A'	'B'	'C'	'D'

Use this instead

DN1 GREATER THAN DN2	Y	–	–	–
DN1 GREATER THAN DN3	–	Y	–	–
DN1 GREATER THAN DN4	–	–	Y	–
DN1 GREATER THAN DN5	–	–	–	Y
Q =	'A'	'B'	'C'	'D'

Figure 5-18 Incorrect and correct way to handle inequalities that have data items as both operands.

function is partially provided by the NOT option of extended entry conditions.

2. Ability to execute some actions in the process of evaluating conditions. This implies some forced ordering of condition evaluation, but the value of this ability may offset the accompanying loss of efficiency.

3. Range handling with ranges limits being dynamically determined at run time.

4. A repeat, or ditto, character has proven useful in eliminating repetitive writing of identical entries across a row. We do not attempt to provide conventions for the ditto here.

SECTION 6

Actions

After consideration of all of the conditions and their interrelationships which go into the selection of a specific rule of the decision table, that rule must do something; namely, execute a series of actions.

ACTION ROWS AS OPERATIONAL STATEMENTS

Actions are operational steps (statements) to be taken in some specified sequence, the precise operational steps and sequence being variable from one rule to another.

Operational steps are imperative procedure statements such as those that follow[1]:

      MULTIPLY GROSS BY RATE GIVING TAX.
or,   MOVE A TO B.
or,   GO TO NEXT-TABLE.

In addition to single statements, entire sentences, groups of sentences or conditional statements may be used, as shown in the following:

    PERFORM MEAN-CALCULATION, ADD DEVIATION TO MEAN, GO TO REPORT-TABLE.
or, GO TO PN1, PN2, PN3 DEPENDING ON VECT.

The preceding example involving the GO TO ... DEPENDING ON ... statement, is a "conditional" statement—that is, the flow of control may be directed to one of several places (in this case, PN1, PN2, PN3, or the next sentence). Conditional

---

[1] COBOL-language decision tables are used in the illustrations; other languages are treated in a similar manner.

40 ACTIONS

statements in action rows of decision tables are discussed in connection with sequences of actions.

Although an operational step may consist of many individual statements, the term "operational statement" is used to mean the more general situation just described as "operation step" and is not meant to limit the textual content of the operational step.

The primary requisite for the text of the operational statement indicated by an action row is that the statements be acceptable members of the host language. In the COBOL case, for example, the text must be such that a period may be added at its end if there is not already one there.

EXPRESSION OF OPERATIONAL STATEMENTS IN AN ACTION ROW

The operational statement must be expressed and associated with one or more rules within any given action row. As with conditions, there are two major ways of doing this—limited and extended entry.

To discuss these two types of entries, it is necessary to recall the basic structure of the decision table, repeated in Figure 6-1.

In the action section, any particular action row contains a stub portion and one or more rule entries (including the ELSE-rule entry, if specified).

Limited Entry Action Row

In the limited entry action row, the entire operational statement is contained in the stub (including any stub continuation procedures which may be available in operational preprocessors).

The rule entry portion of the action row is used to associate the contents of the stub with the rules to which it applies. Usually a single letter "X" serves this purpose, but other conventions, notably action sequence numbers (ASN), may also be used. It should be noted that in limited entry action

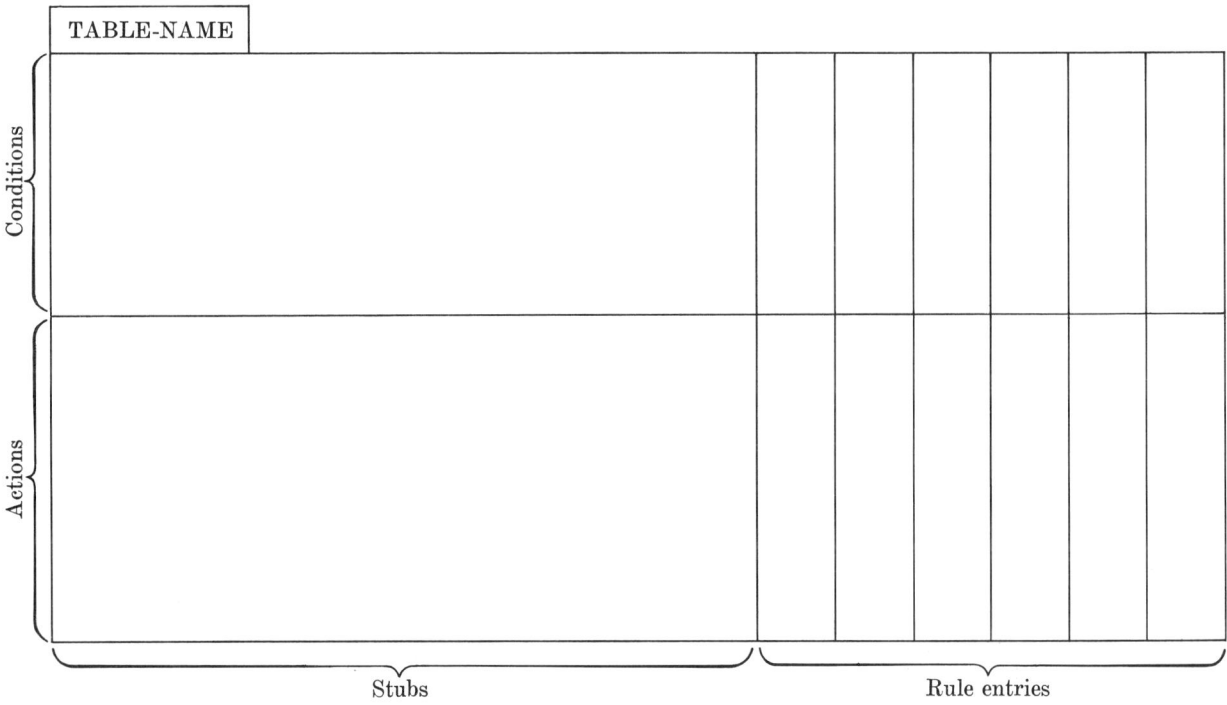

Figure 6-1   Structure of the Decision Table.

	STUB...	RULE ENTRIES...	
C	X = 5	Y	N
A	GO TO Q1	X	–
A	GO TO Q2	–	X

Figure 6-2 Example of limited entry action rows.

rows the contents of the rule entry in no way modifies the operational statement itself but rather merely serves to associate it with the rule, as in Figure 6-2.

Figure 6-2 shows the first action, GO TO Q1, associated with the first rule by an "X." Likewise, GO TO Q2 is associated only with the second rule. The complete operational statement of each is in the stub.

Extended Entry Action Row

In extended entry action rows, the operational statement is only started in the stub with its completion in the rule entry itself. The advantages of this procedure are largely in the decision table's role as a document, rather than as a logical structure. Every extended entry action row is logically equivalent to a set of limited entry action rows as shown in Figure 6-3. The subject of conversion of extended to limited entry is discussed in the section *Conversion of Decision Tables*.

The single action row above the heavy arrow in Figure 6-3 is easier to read, easier to write and more compact than the three limited entry action rows below the arrow.

Extended entry action rows associate the operational statement with various rules by the act of completing the operational statement in the entry portion of desired rule(s). The marker "X" is not used.

Sequence of Action Rows

Actions need to be executed in a particular sequence for each rule. The basic sequence of execution of the operational statements for a given rule is top to bottom. Only those action rows that have something in the rule entry field for the particular rule selected are executed during this top to bottom scan. We use Figure 6-4 to illustrate this.

In Figure 6-4, when the conditions (not shown here) cause Rule 1 to be selected, X is set equal to 1, Y equal to 2, and control transferred to Q in that order. If, instead, Rule 2 is selected, Y is set equal to 2, X equal to 7.2, and control "falls out of" the bottom of the table in that order. In Rule 3, Y is set equal to 2 before the GO TO is executed while X is left unchanged.

		STUB...	RULES...		
Extended Entry:	A	COMPUTE MOD =	A - (A/10)　　3　　C - (C/ENAT)		

⇩

Limited Entry:	A	COMPUTE MOD = A - (A/10)	X		
	A	COMPUTE MOD = 3		X	
	A	COMPUTE MOD = C - (C/ENAT)			X

Figure 6-3 Equivalence of extended-entry action row and its limited-entry action rows.

STUBS...	RULE ENTRIES...			
	1	2	3	4
X = 1	X	–	–	X
Y = 2	X	X	X	–
GO TO Q	X	–	X	X
X =	–	7.2	–	4.3

Figure 6-4 Normal sequence of execution of action operational statements.

Rule 4 might be considered in error. If selected, the top to bottom scan will set X equal to 1 and then transfer to Q. The specified action, set X equal to 4.3, will never be executed since control is transferred out of the table before reaching it.

The "top to bottom" approach may be changed by means of *action sequence numbers (ASN)*. These permit the actions associated with a particular rule to be ordered in any desired way. Conventions for writing ASN's vary but the following convention will be assumed here:

Limited entry: the ASN is used either
1. in lieu of the "X" marker,
or
2. preceding the "X," separated by a right parenthesis.

Extended entry: the ASN is used preceding the rule entry and separated by a right parenthesis.

Figure 6-5 illustrates ASN usage.

In Figure 6-5 Rules 1, 2, and 3 are equivalent—all three set Z equal to 1.15, then Y equal to 2.4, and then GO TO Q1. Rule 4 sets Y, then sets Z before going to Q1. Rule 5 adds 7.22 to Z, first, then computes Y (which is a function of Z and the previous value of Y) and resets Z equal to 1.15 before going to Q2.

ASN's are especially useful when a series of actions are to be executed in different sequences for various rules. In those preprocessors that do not provide for the ASN features, the top to bottom approach requires that the action rows be replicated. This is illustrated in Figure 6-6.

Obvious combinations of the action rows of Figure 6-6 have not been made so that the replication of action rows will be easier to follow.

Whenever ASN's are used, they should be used for all actions in a particular rule. Use of ASN's for some rules while permitting top-to-bottom on different rules in the same table is permitted by most preprocessors that support ASN. Figure 6-7, discussed below, includes an example of such mixed usage for limited entry action rows.

Conditionals in Action Rows

Sometimes a decision is more conveniently made during the actions. Usually all such decisions should be considered as part

STUBS...	RULE ENTRIES...				
	1	2	3	4	5
MOVE 1.15 TO Z	1	1)X	–	2	3)X
MOVE 2.4 TO Y	2	2)X	–	1	–
GO TO Q1	3	3)X	3)X	3	–
COMPUTE Z =	–	–	1)1.15	–	1)7.22 + Z
COMPUTE Y =	–	–	2)2.4	–	2)Z * Y + 2
GO TO	–	–	–	–	4)Q2.

Note: This table (only the action rows are shown) illustrates various ways of using ASN indicators.

Figure 6-5 Illustration of ASN indicators.

EXPRESSION OF OPERATIONAL STATEMENTS IN AN ACTION ROW

STUBS...	RULE ENTRIES...				
	1	2	3	4	5
MOVE 1.15 TO Z	X	X	–	–	–
MOVE 2.4 TO Y	X	X	–	X	–
GO TO Q1	X	X	–	–	–
MOVE 1.15 TO Z	–	–	–	X	
COMPUTE Z =	–	–	1.15	–	7.22 + Z
COMPUTE Y =	–	–	2.4	–	Z * Y + 2
MOVE 1.15 TO Z	–	–	–	–	X
GO TO	–	–	–	–	Q2
GO TO Q1	–	–	X	X	–

Note: The table in Figure 6-5 (conditions not shown) is here rewritten in a top-to-bottom form.

Figure 6-6 Repetition of action statement in lieu of using ASN's.

of the condition section with the definition of new rules but there will be cases in which it is expedient to defer the condition.

Figure 6-7 illustrates just such a case. In the order filling situation presented, it is assumed that the information concerning present stocks of the ordered item (ON-HAND-QTY) is available, along with the quantity ordered (ORDER-QTY). If the supply is inadequate to fill the order, the source of supply must be determined in order to fill or back-order the needed item.

Since locating the vendor of a purchased item may be a time-consuming process, it should be done only after determining the need for it. Normally another table would be entered (see section on *Structures of Tables*), but the use of conditionals in actions, as in Figure 6-7, is often an expedient, if inelegant, alternative.

C	ORDER-QTY > ON-HAND-QTY	N	Y	Y	
C	MFR-IN-HOUSE = 1	–	N	Y	
A	IF BACK-ORDERED-SW = 1 PERFORM BACK-ORDER, ELSE PERFORM FILL-ORDER.	X	4)X	3	
A	SUBTRACT ORDER-QTY FROM ON-HAND-QTY.	X	–	–	
A	PERFORM ISSUE-PURCHASE-OR-WORK-ORDER.	–	3)X	2	
A	IF WORK-ORDER-LEAD-TIME > 14 MOVE 1 TO BACK-ORDERED-SW.	–	–	1	
A	PERFORM LOCATE-SUPPLIER.	–	1)X	–	
A	IF SUPPLIER-LEAD-TIME > 10 MOVE 1 TO BACK-ORDERED-SW.	–	2)X	–	

Note: An example of conditionals used in action rows of a decision table. Various action marker techniques for limited-entry actions are also illustrated. Note that, unlike conditions, the word "IF" must be used to denote this type of conditional.

Figure 6-7 The use of conditionals in action rows.

Special Action Features

A number of Action-related features are either implemented in various real preprocessors or considered as potentially valuable additions. They are mentioned here for completeness.

Interrule Communication

1. Ability to transfer control at a particular action row from one rule to another.
2. Ability to establish dummy rules (rules which cannot be selected directly by the conditions, often called "housekeeping" rules) which may be performed from other rules.
3. Combination of like actions in various rules may be performed as a subroutine. This is done automatically, with various degrees of user control in different preprocessors.
4. Ability to re-execute initialization as an action-row.

Intra-table communications

1. Ability to re-enter the table in a looping fashion from within the table. This may by done by either executing or bypassing initialization.
2. Ability to execute actions while still in the process of evaluating conditions. The mechanism for doing this is difficult to imagine but its value may be appreciated.

Intertable Communication

1. Ability to enter tables from either any action row of other tables or any COBOL coding.
2. Ability to PERFORM tables on the same basis as above.
3. Ability to enter (or start a PERFORM of) a table either through or bypassing initialization.

Other special features of various preprocessors are for purpose of program optimization. Some of these are discussed in Section 7, *Optimization of Decision Tables*.

SECTION 7

Optimization of Decision Tables

There is no such thing as a "most efficient" computer program in general—efficiency is always measured in terms of some resource which is most scarce. In programs the scarce resource is usually either time or machine memory ("core" is referred to hereafter, meaning any type of internal machine memory in which the program is resident as it is executed). One of these two resources—both expressible in terms of money, perhaps the ultimate scarce resource—may be more scarce than the other in any given situation.

In this work we use the word optimization to denote the process that results in minimum core utilization or minimum running time. In decision tables several optimization schemes, often built into the preprocessors, are available to the programmer/analyst so that he may express the criteria to be used in translating (decomposing) the decision table to computer code.

TYPES OF OPTIMIZATION

Preprocessors often offer a choice of either core or run time efficiency for each decision table in an application or system software program.

Core optimization techniques ensure that rules are evaluated in turn, based on the conditions expressed, to minimize the number of source language decisions, that is, minimize the number of IF statements.

"Run time" optimization, the popular name of the other type of optimization, is a minimization of the average logical

46 OPTIMIZATION OF DECISION TABLES

rule selection path length. In this type of optimization, the relative frequency with which the various rules might be hit by a large number of transactions is considered. When processing large volumes of transactions agreeing with this frequency distribution, the total number of IF statements executed is minimized, thus reducing run time.

Optimization Examples

Both of these optimizations operate on the conditions. The logical tree structure represented by the conditions, which leads to the selection of the rules and the execution of the actions associated with those selected rules, might be constructed in several different ways. Each starts off with the same set of conditions and ends up selecting the same rules, but each follows different paths in getting there; for example, the simple decision table in Figure 7-1 (shown

STUBS...		RULE ENTRIES...	
Y = 0	–	Y	N
X = 1	*	Y	Y
X = 2	Y	*	*

Figure 7-1 Decision table to be decomposed.

without its actions) might be represented by any of these three different logical structures, shown as flowcharts in Figures 7-2, -3, and -4.

The term *decomposition* is used to describe any of the techniques by which decision tables are converted into conventional decision trees or programming language code. In the decomposition process, the basic decisions are binary—Yes (Y) or No (N). Whenever a condition is checked, the table is divided into two new tables—those rules in which Y, $, or don't-care entries appear in the condition row just checked are placed

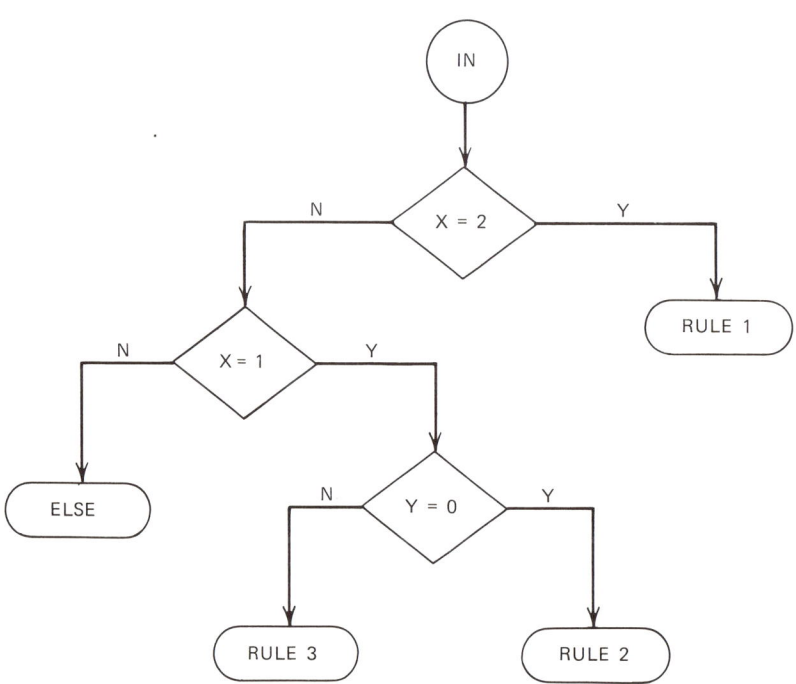

Figure 7-2 Decomposition of table in Figure 7-1, resulting in Run Time optimization; Rule 1 is favored.

TYPES OF OPTIMIZATION 47

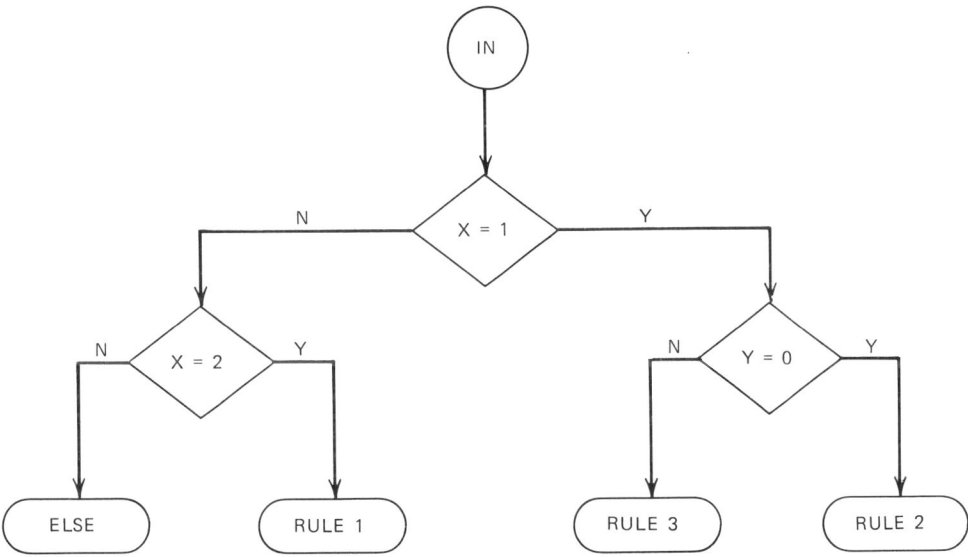

Figure 7-3 Decomposition of table in Figure 7-1, resulting in Run Time optimization; Rule 2 is favored. Note: This decomposition also results in Core Optimization.

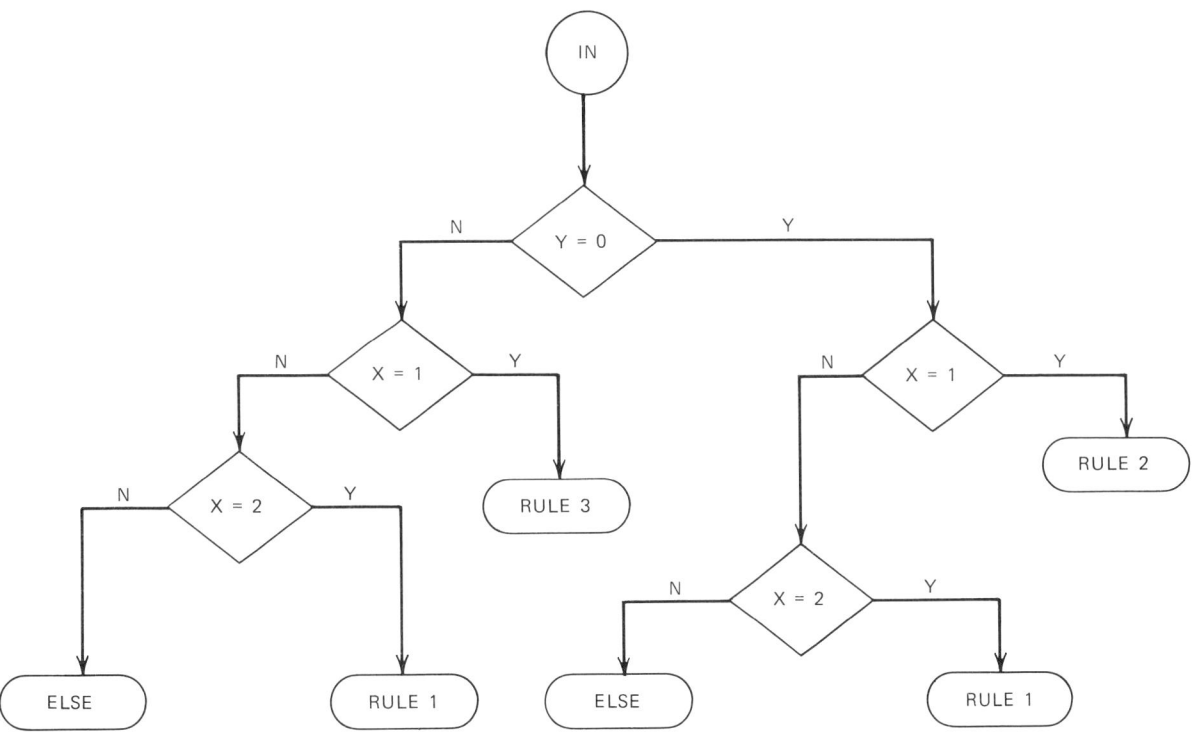

Figure 7-4 Decomposition of table in Figure 7-1, testing conditions from top to bottom.

in one of these "subtables" while those rules in which N, *, or don't-cares appear for that condition are placed in the other subtable. Notice that don't-cares contribute equally to *both* subtables reflecting the fact that don't-cares are, in reality, a combination of Y and N.

The asterisks appear in the table of Figure 7-1 to reflect the mutual exclusiveness of the conditions $X = 1$ and $X = 2$, and may be replaced by N's. Such replacement could produce extra, unnecessary decision steps.

The decomposition in Figure 7-4, which represents a straightforward checking of conditions in the order in which they were written ("top-to-bottom"), has five decision boxes (IF statements), while the decomposition in Figure 7-2, which checked conditions in reverse order, and that of Figure 7-3, which checked the middle condition first, each has three decision boxes.

Run Time Optimization

Run time optimization is based on an expected mix of rule hits by a large number of transactions to be passed against the table. That is, if individual transactions are far more likely to hit certain rules than others, the number of comparisons required to process the total mix should be a minimum based on those expected hits. This implies that those rules most likely to be hit are probably checked for first, although in some cases it is more efficient for the total mix for this not to be the case. Because run time optimization is often not unique, core optimization is sought among those decompositions that satisfy run time optimization. It is not unusual for decompositions based on core optimization and decompositions based on run time optimizations to be the same.

To illustrate, count the number of comparisons needed to identify Rule 1 in Figures 7-2, -3, and -4. In Figure 7-2, Rule 1 is selected after only one comparison, while in Figure 7-3 it is not selected until two comparisons have been made. In Figure 7-4, there are two ways of selecting Rule 1, both requiring three comparisons. If, out of every 1000 transactions to be processed, 900 of

Figure	Rule	Number of Compares	Frequency	Number of Comparisons = Number of Compares Times the Frequency
7-2	1	1	900	900
7-2	2	3	70	210
7-2	3	3	30	90
7-2	(TOTAL)			1200*
7-3	1	2	900	1800
7-3	2	2	70	140
7-3	3	2	30	60
7-3	(TOTAL)			2000*
7-4	1	3	900	2700
7-4	2	2	70	140
7-4	3	2	30	60
7-4	(TOTAL)			2900*

Figure 7-5 Statistics of comparison for Figures 7-2, 3, and 4 when there are 900 hits for Rule 1, 70 hits for Rule 2, and 30 hits for Rule 3.

them expected to hit Rule 1, with 70 hitting Rule 2 and 30 hitting Rule 3, the results obtained are shown in Figure 7-5.

For every batch of transactions that agrees with the expected frequencies, when compared with the Figure 7-2 approach, the Figure 7-4 approach takes 2.4 times as long and the Figure 7-3 approach takes 1.66 times as long. Thus Figure 7-2 is *optimized* for run time given the stated high occurence of Rule 1 hits. But, if we switch the occurrence of 900 hits to Rule 2 and the occurrences of 70 to Rule 1 and 30 to Rule 3, an analysis such as that made above is shown in Figure 7-6.

Notice that Figure 7-3 now does the most efficient job. Figure 7-2 now takes about 1.4 times as long as Figure 7-3, while Figure 7-4, which made such a poor showing on the previous frequency distribution, just barely trails, being 1.035 times the Figure 7-3 time. Thus Figure 7-3 is optimized for run time given a high occurrence of Rule 2 hits.

Clearly, the relative frequency with which rules might be hit is an important factor in the work of the preprocessor and most permit some manner of specifying these relative frequencies. In all of the above, we have assumed equal processing time for each of the conditions. For a discussion of run time optimization that takes this factor into account, see the section *Conversion of Decision Tables*.

Core Usage Optimization

Core usage optimization can be thought of as Run Time optimization, given equal hit probabilities for all rules. In the example case just discussed, Figure 7-3 will be found to be optimized for equal hit probability of the rules and thus is also optimized for core usage.

The process of optimization directly ties into that of decomposition of a decision table and is handled by elaborate algorithms. The user need not be involved with the mechanics of optimization, but should know the effects so that he may use its capability to the greatest advantage.

Optimization of Actions

Optimization of the condition checking procedure that we describe above is not the

Figure	Rule	Number of Compares	Frequency	Number of Comparisons = Number of Compares Times the Frequency
7-2	1	1	70	70
7-2	2	3	900	2700
7-2	3	3	30	90
7-2	(TOTAL)			2860*
7-3	1	2	70	140
7-3	2	2	900	1800
7-3	3	2	30	60
7-3	(TOTAL)			2000*
7-4	1	3	70	210
7-4	2	2	900	1800
7-4	3	2	30	60
7-4	(TOTAL)			2070*

Figure 7-6 Statistics of comparisons for Figures 7-2, 3, and 4 when there are 70 hits for Rule 1, 900 for Rule 2, and 30 for Rule 3.

50 OPTIMIZATION OF DECISION TABLES

only kind of optimization offered by most decision table preprocessors. The action portion of decision tables is also subject to optimization.

Action optimization consists of grouping together like actions and having the rules either GO TO the grouped actions (in which all of the actions of two or more rules are alike) or PERFORM a series of actions which are placed out-of-line (for like sequences within the actions of two or more rules).

Manual Optimization

Since the preprocessor puts out a source language program, COBOL in the case of most commercial language preprocessors, the programmer has the choice of considering this source language as a "starting point" for his own work, modifying it for his own purposes. We now discuss two areas in which the programmer may optimize his program; of course, as preprocessors improve, these and other areas may become incorporated.

When two rules have identical strings of actions and vary in only one condition,[1] the two rules may be combined using the combination rules shown in Figure 7-7 for that one condition's different entries. It doesn't matter which rule comes first. Those marked "possible error" in Figure 7-7 are combinations that would not occur in a table that was perfectly correct and which also made full use of the * and $ entries at every occasion. These possible error combinations should be investigated whenever they are encountered.

Rule combination are more fully discussed in the section *Rules: Completeness and Rule Combination.*

---

[1] We consider * and N to be the same and Y and $ to be the same for this purpose only.

Entry a	Entry b	Combined Rule Entry
– (don't care)	anything	–
Y	N	–
Y	*	– (possible error)
Y	$	Y (possible error)
N	*	N (possible error)
N	$	– (possible error)
$	*	– (possible error)

Figure 7-7 Table of combined rule entries.

COMBINING DUPLICATE SECTIONS OF THE RULE SELECTION STRUCTURE

The second area in which the programmer may "postprocess" the source language output of the preprocessor is that of combining like portions of the logical rule selection structure.

Eventually in many logical structures, different branches of the structure reach a point where they make identical checks (decision boxes) with identical Yes and No results. To illustrate, in the earlier example, Figure 7-4 ("top to bottom table analysis"), the two $X = 2$ decision boxes fall in this category and, thus, the two decisions may be combined. To perform this combination simply have the GO TO generated to one of the IF $X = 2$ ... statements go instead to the other one, eliminating the first. Although Figure 7-4 does not represent either a core or run time optimization, such redundant structures could occur even within properly optimized structures.

As a final note on optimization it should be pointed out (as illustrated in Figures 7-2, -3 and -4) that the order of the condition rows themselves have little to do with process of optimization. It is this fact—that the conditions might be checked in any order rather than the one in which they were written—that accounts for the assumed nonrelatedness of the conditions themselves. Consequently, special techniques must be used to express any relationship that does exist between conditions, as is discussed in the earlier section on conditions. One cannot rely on the order of the condition rows themselves to express any meaningful ordering.

SECTION 8

The Rule: Redundancy and Contradiction

Recall that one of the basic assumptions for decision tables is that for each table the selection of a rule is unique. Any possible transaction that might be passed against that table will select one and only one rule (assuming the ELSE-rule to be present). This requirement greatly affects the use of decision tables; we now consider some of these effects.

The techniques discussed in this chapter are aimed primarily at decision tables for computer programs. However, they are equally applicable and useful for checking out tables used for system analysis and for instructional purposes.

RULE UNIQUENESS

The requirement for uniqueness of rule selection is the primary constraint in the writing of a decision table. For any pair of rules, the various conditions must be connected together in such a way that there exist mutually exclusive entries on at least one condition for that pair of rules. Consider the two conditions in Figure 8-1.

The first condition has a mutually exclusive entry between Rules 1 and 2, Rules 1 and 3, Rules 2 and 4, as well as between Rules 3 and 4. The second condition has mutually exclusive entries between Rules 2 and 3, as well as between Rules 2 and 4. No mutually exclusive entry exists for the first or second condition of Rules 1 and 4. Thus any transaction passed against the table that has Q equal to 1 and P equal to 2 will cause both Rules 1 and 4 to be selected. This is in violation of

STUBS...	RULE ENTRIES...			
	1	2	3	4
Q EQUAL TO 1	Y	N	N	Y
P EQUAL TO 2	–	N	Y	Y

Figure 8-1 Conditions to illustrate mutual exclusion and nonexclusion.

the basic requirement of unique rule selection in a decision table.[1]

CONTRADICTION AND REDUNDANCY

The absence of mutually exclusive entries for at least one condition between two given rules implies that some transaction can be found that satisfies both rules, as is illustrated in Figure 8-1. Whether this situation is a redundancy of rules or a contradiction of rules depends on whether or not the series of actions for each of the two rules are the same.

If the series of actions are the same, redundancy exists. Inasmuch as both rules lead to the same series of actions anyway, it really doesn't matter that they are both selected.

If the series of actions is different, however, the possibility of executing two different strings of actions for the particular transaction arises; this is, in fact, a contradiction.

To illustrate, consider the tables in Figures 8-2, -3, and -4.

In Figures 8-2 and -3, neither condition has mutually exclusive entries between the two rules. In both Figures 8-2 and -3, a transaction such as one which has P equal

---

[1] It is this mutual exclusiveness of rules that historically gave rise to the *Traditional Decision Table* requirement that every entry of any given condition be either mutually exclusive or identical to each of the other entries for that same condition, excepting don't-cares.

STUBS...	RULE ENTRIES...	
C P = 1	N	N
C Q = 2	–	N
A GO TO	R	S

Figure 8-2 Contradiction.

STUBS...	RULE ENTRIES...	
C P = 1	N	N
C Q = 2	–	N
A GO TO	R	R

Figure 8-3 Redundancy.

STUBS...	RULE ENTRIES...	
C P = 1	Y	N
C Q = 2	–	N
A GO TO	R	S

Figure 8-4 Correct table.

to 3 and Q equal to 4 would cause both rules to be selected. In the case of Figure 8-2, different actions (GO TO R and GO TO S) are specified, resulting in two contradictory decision rules. In case 3, however, both rules have the same action (GO TO R), resulting in redundancy.

The table of Figure 8-4 is a comparable table without redundancy or contradiction, since the first condition cannot be satisfied for both rules at once, that is, for any possible transactions either P equals 1 or P does not equal 1, but not both.

Two rules which are neither redundant nor contradictory are said to be *independent* of one another. Note that this usage is quite different from the way the word *dependent* is used with conditions.

A complete discussion on manual checking procedures for determining redundancies and contradictions is discussed in the section *Writing the Decision Table*.

SECTION 9

The Rule: Completeness and Rule Combination

SIMPLE RULES

The Simple Rule is a rule that has no don't-cares; that is, the rule contains a Y, N, *, $ or an extended entry for every condition of the table.

Condition entries may be grouped, for simple-rule purposes, into two groups: Y and $ in one group; and N and * in another. Extended entries, as are discussed more completely in the later section on *Algorithms for Automatic Translation*, are convertible into one of these four entries for any given rule. An exception is the bound-operator extended entry rows whose rule entries may not be considered as a member of either group. Bound-operator conditions are excluded from this chapter.

These two groups, the one containing Y and $ and the one containing N and *, are mutually exclusive. Thus, a non-don't-care condition entry might assume two states: *true* (Y or $) or *false* (N or *).

Excluding for the moment extended entry conditions, if there are n conditions in the table, there are 2^n independent simple rules possible for that table. The "2" in this function represents the number of states the entry may take. If there are two limited entry conditions in the table, there are no more than 4 ($=2^2$) simple independent rules possible without contradiction or redundancy. Figure 9-1 shows such a case.

BINARY MATRIX TABLES

We denote the table shown in Figure 9-1 as a *binary matrix table*. A binary matrix table is one that shows every possible combination of Y and N for the conditions present.

STUBS...		RULE ENTRIES...			
C	P = 1	N	N	Y	Y
C	Q = 2	N	Y	N	Y
A	GO TO	R	R	R	S

Figure 9-1 Table of four independent rules.

Notice that every pair of rules in this table has at least one condition whose entries for the two rules are mutually exclusive. Thus there is no redundancy or contradiction. Further, notice that no other rule can be added which will share this property of mutual exclusiveness. Hence, following the 2^n, four is the maximum number of rules that can be written for a two condition table; for a three condition table, up to eight, but no more than that, could be written; and so forth.

The presence of a $ or * entry implies that only that state is possible. Thus the 2^n procedure does not accurately reflect the number of "practically possible" simple rules.

EXTENDED ENTRIES AND THE BINARY MATRIX

For simplicity's sake, we excluded extended entries from the original discussion of the binary matrix, defining the binary matrix for limited entry type tables. But some of the implications of the binary table to be discussed shortly are important enough to require a broadening of the discussion to include extended entry.

Ruling out the special bound-operator extended entry rows, and any other entry types which might include don't-cares in their expansions, we can easily calculate how an extended entry affects the number of simple rules possible. This effect is related to the technique by which the extended entry condition is expanded into a set of limited entry conditions during the conversion of the decision table to program code, a topic discussed in a later section. This conversion varies according to the type of extended entry but a procedure can be derived which holds for all the various conversions. That procedure is to count each different entry that the extended condition has over all of the rules and consider the extended condition to represent that many limited conditions. Each of these limited conditions will have some non-don't-care entry for each rule that had a non-don't-care extended entry in the original entry. The n in the 2^n functions representing the number of simple rules possible will include these assumed limited entry conditions. The table shown in Figure 9-2 illustrates this.

The condition shown in Figure 9-2 would be counted as three conditions when calculating the number of simple rules possible, since it has three different non-don't-care entries: 17, 4, and 12. This extended condition's contribution to n is three. Non-overlapping bounded ranges should be considered in terms of the ranges involved, rather than the conditions that make up these ranges. Several examples illustrate this later in this section.

COMBINING RULES

Returning to Figure 9-1, the first, second, and third rules all have the same action. Since each of the first three rules end up executing the same action, it may appear that those rules could be combined into a single rule. But all three cannot be combined. However, two of them can be as shown in Figure 9-3.

Any two rules with identical actions and which differ in only one condition entry—a "Y" and an "N"—may be combined. The differing entry becomes a don't-care ("–"). (Note, for purposes of combination, explicit and implicit entries of the

COMBINING RULES 57

STUBS...	RULE ENTRIES...						
EXT-ENT =	17	4	12	–	17	4	–

Although $n = 3$ (there are three values: 4, 12, and 17), one would expect there to be $2^3 = 8$ independent simple rules. In fact, because of the condition dependence, there are only four.

	R1	R2	R3	R4
EXT-ENT = 4	Y	*	*	N
EXT-ENT = 12	*	Y	*	N
EXT-ENT = 17	*	*	Y	N

The other four rules can be written, but they obviously are impossible rules.

	R5	R6	R7	R8
EXT-ENT = 4	Y	Y	Y	*
EXT-ENT = 12	Y	Y	*	Y
EXT-ENT = 17	Y	*	Y	Y

Figure 9-2 An extended entry condition.

	STUBS...	RULE ENTRIES...		
C	P = 1	–	N	Y
C	Q = 2	N	Y	Y
A	GO TO	R	R	S

Figure 9-3 Combining Rules 1 and 3 of Figure 9-1.

same sense are considered as being equivalent, i.e., * and N are considered equivalent, as are $ and Y.) The second and third rules in Figure 9-1 cannot be combined since they differ in two conditions.

Because a rule with r dashes is equivalent to 2^r simple rules, it must be apparent that three (which is not a power of 2) simple rules cannot be combined into one rule. However, three complex rules (having the same series of actions, of course) may be combined into two rules, if the three complex rules are equivalent to 2^x simple rules; for example, in Figure 9-4 the first rule of the leftmost three rules can be decomposed into two simple rules so that the three rules become the following four rules.

STUBS ...	RULES ...			
	1	2	3	4
C1	N	N	Y	N
C2	Y	N	Y	N
C3	N	N	N	Y

Rules 1 and 3 and Rules 2 and 4 can then be combined to form the two rightmost rules of Figure 9-4.

This example is provided to show how three rules were converted to two. It is not suggested that decision table writers attempt to do this type of combining. The occurrences of such convertible combinations of rules is small enough to not warrant the effort involved in reducing the number of rules in the table.

STUBS...	RULE ENTRIES...				STUBS...	RULE ENTRIES...	
C1	N	Y	N		C1	–	N
C2	–	Y	N	*is equivalent to*	C2	Y	N
C3	N	N	Y		C3	N	–

Figure 9-4 Conversion of three rules (simple and complex) to two complex rules.

General Requirements for Combining Rules

In general, if any two rules have the same series of actions and are alike (i.e., of the same sense) in every condition but one, (or in the cases of bounded ranges, differ only in one range definition) and in that one difference the entry for one of the rules is the negation of the entry for the other rule (or, in the case of bounded ranges, their ranges are contiguous), then those two rules may be combined into a single rule.

If the only difference between two rules is an extended (but not bounded range) entry, the two rules usually cannot be combined since one entry will not usually be the negation of the other.

Rules for combination of condition entries are discussed and presented as a table in the section on *Optimization of Decision Tables*. Combination of contiguous ranges is discussed later in this section. To illustrate combination of rules with limited entries, consider Figures 9-5 and 9-6 before and after combinations.

Notice that the "combination of rules" being discussed is for the combining of independent rules, that is, rules that do not contain redundancy between them.

STUBS...			RULE ENTRIES...
C	P	EQUAL TO 1	Y Y
C	Q	EQUAL TO 2	N Y
A		GO TO R.	X X

Figure 9-5 Decision table before combining rules.

STUBS...			RULE ENTRIES...
C	P	EQUAL TO 1	Y
C	Q	EQUAL TO 2	–
A		GO TO R.	X

Figure 9-6 Decision table after combining rules.

"COMPOUND" AND "AS WRITTEN" RULES

An important point in Figure 9-6 is that the combined rule is a complex rule representing the two simple rules of Figure 9-5. Whenever a don't-care appears in the condition of a compound rule, it doubles the number of simple rules represented by that compound rule. Excluding, for a moment, extended entries, if there are k dashes (or blanks) in the conditions of a compound rule, then that rule contains $r = 2^k$ simple rules. The terms *rule* and *as written rule* without the qualifier *simple* may mean either type. The rule *as written* in an actual decision table may be either a simple or a compound rule.

COMPLETENESS CHECKING

Summing the number of simple rules represented by the individual rules ($rc_j' = 2^k$) gives the count of simple rules represented by the full table ($\Sigma rc_j'$). Comparing this count to rc, the rule count derived from the number of conditions n ($rc = 2^n$) produces a measure of the "completeness" of the table. The sum of simple rules present in the table is illustrated in Figure 9-7.

Completeness checking answers the always present question of "Did I allow for every case in my decision table?" Any case not covered by the written rules will of course be covered by the ELSE-rule. Completeness checking is also dependent on every condition pertinent to the problem under consideration having been expressed. If, in summing up the count of simple rules in your table, you reach a total that exceeds the maximum 2^n, then a contradiction or redundancy exists somewhere in your table.

Completeness Checking Involving Extended and Bounded Range Conditions

In the discussion of the binary matrix earlier, the role of extended entries is de-

STUBS...	RULE ENTRIES...					NUMBER OF LIMITED ENTRY CONDITIONS REPRESENTED
A = 1	Y	N	–	–	N	1
B = 2	–	Y	–	Y	N	1
C = 3	–	Y	N	N	N	1
D = 4	Y	–	N	Y	Y	1
E = 5	N	–	–	Y	Y	<u>1</u>
Number of dashes = k	2	2	3	1	0	n = <u>5</u>
Number of simple rules represented = (2^k)	(2^2) = <u>4</u>	(2^2) = <u>4</u>	(2^3) = <u>8</u>	(2^1) = <u>2</u>	(2^0) = <u>1</u>	
CALCULATION $rc = 2^n = 2^5 = 32$ simple rules possible $rc' = \Sigma\, 2^k = (4 + 4 + 8 + 2 + 1) = 19$ Simple rules missing $= (rc\text{-}rc')$: $(rc - rc') = (32 - 19) = 13$ simple rules represented by an ELSE-rule						

Figure 9-7 Completeness checking in a decision table with limited entry condition rows.

scribed. A similar situation exists in completeness checking.

Every unique value in any extended condition, regardless of how often it may be used in that condition row, can be thought of as a limited entry condition to itself, with whatever don't-cares from the original extended entry carried over to the limited entry row. Each of these limited entry conditions is to be added into the condition count, n, and don't-cares reflected in the limited entries are to be added into the don't-care count, k, for each rule. Nonoverlapping ranges are counted as *number of ranges* towards n, and as *number of ranges* toward k. The equivalent handling of overlapping ranges is not covered in this book.

An example of completion checking of a table containing normal extended and bounded-ranges extended conditions is shown in Figure 9-8.

THE ELSE-RULE

The difference, $rc - rc'$, has been shown to express the number of simple rules not covered by the as-written rules. The decision table, upon encountering a transaction agreeing with any of these missing simple rules, will select the ELSE-rule. The ELSE-rule handles all of those transactions that are not allowed for in the as-written rules of the decision table.

If $rc - rc'$ equals zero, as it does for a binary matrix, no transactions will be directed to the ELSE-rule. But if rc' is less than rc it is possible for some transaction to be directed to the ELSE-rule. The case of rc' greater than rc, as has been mentioned, indicates a redundancy or contradiction in the table.

Unless the decision table writer has made a completion check and found perfect

STUBS...	RULE ENTRIES...					NUMBER OF LIMITED ENTRY CONDITIONS REPRESENTED
LEVELS GT	–	10	22	22	30	3
LEVELS LE	–	22	30	30	45	
POLICY-TYPE =	'A'	'B'	'C'	'D'	–	4
RIDER = 'X'	Y	–	Y	–	N	1
Number of limited entry conditions represented by dash = k Number of simple rules represented = (2^k)	3 (2^3) = $\underline{8}$	1 (2^1) = $\underline{2}$	0 (2^0) = $\underline{1}$	1 (2^1) = $\underline{2}$	4 (2^4) = $\underline{16}$	$n = \underline{8}$
CALCULATION $rc = 2^n = 2^8 = 256$ simple rules possible $rc' = \Sigma\, 2^k = (8 + 2 + 1 + 2 + 16) = 29$ Simple rules missing = $(rc\text{-}rc')$: $(rc - rc') = (256 - 29) = 227$ simple rules represented by an ELSE-rule						

Figure 9-8 Completeness checking in a decision table with limited entry and extended entry condition rows.

completion (rc equal to rc'), an ELSE-rule should always be provided. If any extended entry conditions appear in the table, an ELSE-rule is advisable, even if the table appears complete. If not provided, and a transaction is directed to it, the particular preprocessor's default ELSE-rule (if any) will usually take some action, such as stopping the program operation.

Usual and common practice, as well as good control, suggests that every case possible—including *possible* error situations—be covered by the as-written rules of the decision table. This leaves only the *impossible* cases (i.e., cases in which a validation check is made, or logically impossible) to be directed to the ELSE-rule.

The discipline of thinking through every possible combination of the conditions to as great an extent as possible will often pay off to the analyst and programmer. The ELSE-rule should be written as an error procedure. The use of the ELSE-rule as a residual rule is discouraged.

COMBINATIONS INVOLVING IMPLIED CONDITION ENTRIES

Handling of the implicit * and $ entries during combination of rules is rather involved, for they represent condition dependencies that may exist in one simple rule and not in the other with which it is being combined. In most cases, if an implicit * or $ is involved with an explicit N or Y entry in a rule combination step, an error has been made earlier. Most likely a Y or N had been used in some rule where an * or $ would have served as well. The safest approach to

COMBINATIONS INVOLVING IMPLIED CONDITION ENTRIES

STUBS...		RULE ENTRIES...				
BEFORE		1	2	3	4	5
C	CLASSX = 'A'	Y	–	*	*	Y
C	CLASSX = 'N'	*	–	Y	Y	*
C	SIZEX > 0	($)	Y	$	$	(Y)
C	SIZEX > 5	Y	–	$	$	Y
C	SIZEX > 10	–	–	Y	Y	–
C	USAGEX =	'D'	–	'D'	'C'	'D'
C	USAGEX = 'F'	*	Y	*	*	*
A	GØ TØ	R.	T.	S.	S.	R.
AFTER						
C	CLASSX = 'A'	Y	–	*	*	
C	CLASSX = 'N'	*	–	Y	Y	
C	SIZEX > 0	(Y)	Y	$	$	
C	SIZEX > 5	Y	–	$	$	
C	SIZEX > 10	–	–	Y	Y	
C	USAGEX =	'D'	–	'D'	'C'	
C	USAGEX = 'F'	*	Y	*	*	
A	GØ TØ	R.	T.	S.	S.	

Figure 9-9 Combining rules containing implied condition entries. (A) Since all conditions but one are equal in rules 1 and 5 and all actions are equal, they become candidates for combination. The Y and $ are combined into a Y. (B) Rules 3 and 4 cannot be combined due to the different normal extended entries. (C) If X > 5, obviously X must also be greater than zero, so the circled "Y" at C may be changed back to a "$."

this matter is to change the * to an N, the $ to a Y, and then combine normally. The combined result might be re-evaluated to see if the dependency still exists and if so, the * or $ may be reinstated. Figure 9-9 includes an example of such combination.

Combination of rules involving bounded extended entries, especially the NE operator case, had best be approached only after some experience with combining other types of rules and working with bounded extended entries. If two ranges are contiguous (with the equal case between them included) the ranges may be combined. Various combination situations are shown in Figure 9-9 ($ and normal extended entry) and Figure 9-10 (bounded extended entry).

As a final note on combination, it should be pointed out that some preprocessors do some rule combination in the course of developing the generated computer code. Combining rules may not be advantageous from an efficiency point of view, as for instance, when reduction in computer running time is desired, assuming that the relative frequency of occurrence of different transaction types is known. Also, the role of the decision table as a documentation aid should be considered when the possibility for combining occurs.

62 THE RULE: COMPLETENESS AND RULE COMBINATION

	STUBS...	RULE ENTRIES...				
	BEFORE	1	2	3	4	5
C	PROC-NM =	'A'	'A'	'N'	'N'	'N'
C	LNGTH GT	0	8	–	–	–
C	LNGTH LE	8	30	–	–	–
C	LNGTH GT	–	–	0	10	–
C	LNGTH LT	–	–	10	19	–
C	LNGTH = 10	–	–	*	*	Y
A	PERFORM	X1.	X1.	X3.	X3.	X4.
		(A)		(B)		
	AFTER					
C	PROC-NM =	'A'	'N'	'N'	'N'	
C	LNGTH GT	0	–	–	–	
C	LNGTH LE	30	–	–	–	
C	LNGTH GT	–	0	10	–	
C	LNGTH LT	–	10	19	–	
C	LNGTH = 10	–	*	*	Y	
A	PERFORM	X1.	X3.	X3.	X4.	

Note: Some simplification could be done to eliminate the number of conditions in this table by proper analysis of the conditions.

Figure 9-10 Combining rules involving bounded extended entry. *Contiguous ranges:* (A) The two ranges [> 0 through 8], and [> 8 through 30] are contiguous and constitute the only differences between first two rules (which have identical action series). Hence, these rules may be combined. Notice that TWO condition rows differ between those rules but when interpreted in terms of ranges, there is only one difference. *Noncontiguous ranges:* (B) The ranges are the only differences between these two rules, but since neither limit contains the equal case between them, the two ranges are not contiguous and thus may not be combined.

SECTION 10

The Languages Used in Decision Tables

It has been mentioned previously that a variety of languages are used in decision tables: COBOL, FORTRAN, ALGOL, or any number of other languages including the natural languages such as English.

For decision tables used in computer programs, each particular decision table preprocessor will specify the language or languages that may be used. The manual or user's guide for a specific preprocessor will explain restrictions or extensions to the particular language which are of interest to their users.

The three most commonly used languages—English, COBOL and FORTRAN—are briefly discussed here. Each of these languages has a large vocabulary and rules of usage that are beyond the scope of this section.

SYNTAX AND SEMANTICS

Each of our languages (English, COBOL, and FORTRAN) may be considered as a combination of two parts: *syntax* and *semantics*.

Syntax

Syntax is the structure of a language. To illustrate, in COBOL the structure (i.e., syntax) of the MOVE statement (simplified) is,

$$MOVE \begin{Bmatrix} \text{data-name-1} \\ \text{literal} \end{Bmatrix} TO \text{ data-name-2 [data-name-3] } \ldots$$

This structure means that after the word MOVE, only a data-name or a literal may appear (the two are in braces, indicating a choice). After this, only the word TO may appear. Following TO, a data-name *must* appear (data-name-2 is not in brackets indicating that it is required), and any number of data-names *may* follow that required one (being in brackets, data-name-3 is optional and the ellipsis ("...") following indicates it may be repeated.

As allowed by the syntax, the following are legal COBOL MOVE statements:

MOVE JOE TO MARY
MOVE MARY TO JOE
MOVE 30 TO TOM DICK HARRY

But the following are not legal, because they violate the syntax (or structure) for the MOVE statement as it was stated above:

MOVE JOE MARY TO TOWN (JOE MARY is neither a data-name nor a literal.)
JOE MOVE TO MARY
JOE TO MARY MOVE

To sum up, syntax has to do with the allowed order of words in a meaningful statement.

Semantics

Words, of course, are not simply strings of letters. They have meanings and the exact meaning of a word in a statement is the concern of semantics.

In the MOVE example given in the discussion of syntax, the names JOE, MARY and others were used, in a syntactically correct manner, as data-names. If, in the COBOL program that contains "MOVE JOE TO MARY", JOE is defined as procedure-name rather than as a data-name, then the statement is incorrect because of semantics.

LANGUAGES

Both the syntax and semantics of any natural language are very involved and often not fully definable in any formal way. The *grammar* of a natural language includes both the areas of syntax and semantics.

Programming languages are artificial, rather than natural, languages. They are modeled after both the natural language and after the symbolic notation which might be called the "natural language" of mathematics. COBOL tends much more to English-like syntax while FORTRAN is closer to being mathematics-like. All programming languages were designed to aid and simplify the statement of a problem in terms that are easily translated to machine-executable instructions.

English

English is widely used in decision tables. System analysts often use English freely or define a special subset related to their problem. An example from a manufacturing system analysis is shown in Figure 10-1.

One major semantic question here is, what does *fit* mean? It could be interpreted as "inside diameter of the bearing equal to outside diameter of the shaft," or perhaps "does the threaded mounting parts of each have the same thread characteristics?" or a thousand other interpretations.

The English-language statement often relies on the reader "knowing what is meant," within a particular context. Within the particular manufacturing operation,

	STUBS...	RULES...	
C	Will this bearing fit that shaft?	Y	N
A	Use this bearing	X	–
A	Try another bearing	–	X

Figure 10-1 Sample Decision Table Containing English.

whose analysis included the table in Figure 10-1, the meaning will be clear. Everyone involved will "know what fit means" so there would be no point in explaining matters further.

In attempting to analyze a system, or organize some information or procedure that is not intended to be directly converted to a computer program, or simply to help organize one's own thoughts, the English language table is exceptionally well suited for use.

COBOL

COBOL is a procedural programming language that is used to define, step by step, operations to be followed in performing systems tasks. Commercial work is its primary application; indeed the name COBOL is derived from the phrase COmmon Business Oriented Language. Strong in the areas of logical decision making itself, COBOL is perhaps the favorite language utilized in decision tables that are part of computer programs. Appendix II, *An Introduction to COBOL*, is available for those who are not COBOL programmers.

COBOL'S basic logical structure is the IF statement whose syntax (simplified) is,

IF condition THEN true-action [ELSE false-action].

The condition might be the equality of the value of two data items (IF X = Y . . .) or any number of other forms. Since COBOL language programs must be translated to machine executable instructions by a computer program known as a compiler, the full set of syntactical and semantical requirements can be obtained from the manual for the particular COBOL compiler to be used. COBOL has been standardized by ANSI, the American National Standards Institute (formerly USASI). COBOL compilers conforming to ANSI standards are called ANS COBOL compilers.

To return to the IF statement, if a condition is evaluated and found true, then the COBOL statements in the *true-action* are executed. If the condition is found false, those statements in the optional *false-action* are executed. This is similar to the mutual exclusion between rules within a single limited entry condition row of a decision table.

COBOL-oriented decision tables are often decomposed into a series of IF statements linked to one another, and to the table actions by GO TO statements. GO TO's direct the sequence of execution of the program, to permit the full checking of all conditions required to select the one rule of the decision table whose actions will be executed. Different preprocessors often generate logically similar, but physically different, coding. These differences result from the differences in the algorithms used to achieve efficiencies. One procedure for setting up this structure of IF statements is discussed under the subject of Decomposition.

Another COBOL statement of interest in connection with decision tables is the PERFORM statement which permits execution of *subroutines* that are located outside of the table or execution of other tables. Numerous other COBOL statements, including debugging aids, may be studied by those interested in COBOL-oriented decision tables.

FORTRAN

This language is known as a scientific and engineering applications language but it is also used for commercial programming.

Two types of IF statements occur in FORTRAN. The older is a simple comparison of two operands with three actions: greater, equal, or less. The newer form is a two-way compare and permits a greater variety of types of comparisons than does the older form. Its syntax (simplified) is,

<u>IF</u> (operand-1 .xx. operand-2) THEN true-action.

.xx. represents a number of logical operators such as .EQ., .LT., etc. These FORTRAN operators should not be confused with the bound-operators of decision table conditions despite the use of the same set of operator names. The above FORTRAN comparison form, like the COBOL form, is a binary (two-way) comparison whose results are mutually exclusive.

FORTRAN also contains other operators of interest to decision table users, such as GO TO and DO. DO causes repetitive execution of a series of statements following it with control over a variable. The DO is somewhat similar to the COBOL PERFORM, but the "assigned GO TO" feature is popularly used for remote subroutine execution.

FORTRAN compilers have been provided for almost every major general-purpose computer system on the market. It has also been standardized by ANSI.

Other Languages

Almost any programming language has been or could be used in conjunction with decision tables.

These include ALGOL, PL-I, BASIC, and assembly languages. New programming systems are constantly emanating from computer manufacturers, software companies, universities, and professional and industry organizations; standardization of these languages is an active area. All of these newer languages have provision for making binary decisions (e. g., IF) and for altering the flow of control (e. g., GO TO). Both of these are basic elements in the decomposition of decision tables to computer code.

COBOL Macros

COBOL is known as a *higher level language* because of its ability to represent commonly used operations and procedures in a more concise, and consequently much shorter, way than is possible using assembly language. Thus a single COBOL (or FORTRAN, or any other higher level language for that matter) statement is more encompassing than a single assembly language statement. COBOL can therefore be considered a *macro* form of assembly language.

However, certain COBOL statements are quite commonly used together; thus they may be considered as eligible for being grouped and represented by new, even higher level, macros. One good example of this is IBM's COBOL-F compiler (for the S/370 line, OS operating system) in its treatment of the Report Writer features. The compiler internally preprocesses such Report Writer statements as "GENERATE" into other COBOL code such as WRITE statements that are then compiled as if the programmer had originally written such statements. Inasmuch as these Report Writer statements are part of ANS COBOL, the term *COBOL macro* is not really appropriate to them, but they do illustrate the nature of a macro representation of a high-level language.

Macros can be defined as statements that represent a larger body of statements of a lower level language.

Macros can be either parametric or nonparametric. That is, variables may be provided which will be substituted into the macros at specified places or change the macro's structure if they are parametric, but no substitutions are permitted in nonparametric macros. One special case of nonparametric macros, known as *shorthand*, is widely used by commercial programmers to cut down on the amount of writing required in COBOL programming. Shorthand usually does just what its name implies—it serves primarily to reduce writing. The COBOL standard itself provides a limited number of shorthands (also called "abbreviations") in its Report Writer feature.

	STUBS...	RULES...					
C	SUPPLY-OK (SHAFT, BEARING)	Y	N	N	N	N	N
C	GET-FROM (IN-HOUSE)	–	Y	Y	N	N	N
C	GET-FROM (VENDOR)	–	–	–	Y	Y	N
C	AVAILABLE (REQUIRE)	–	Y	N	Y	N	–
A	ITEM-ORDER (BEARING)	X	X	X	X	X	X
A	ORDER (FILL)	X	–	–	–	–	–
A	ORDER (MODEL-SHOP)	–	–	–	–	X	X
A	ORDER (BACK, WORK)	–	X	X	–	–	–
A	ORDER (BACK, PURCH)	–	–	–	X	–	–
A	QUALIFY (PRIORITY)	–	–	X	–	X	X

Figure 10-2 Decision table with macros.

Parametric macros permit the programmer, when using them, to indicate data names to be substituted into the structure. More advanced macros not only allow the simple substitution of names but can also restructure themselves, sometimes in drastic ways, depending on the parameters supplied. The details of this type of macro are well

	STUBS...	RULES...					
C	QUANT OF SHAFT OF TRANSACTION NOT > ON-HAND OF BEARING OF MASTER-FILE-RECORD	Y	N	N	N	N	N
C	SUPPLIER OF MASTER-FILE-RECORD = 'F' OR = 'W' OR = 'G' OR PRODUCTION-CURRENT = 'Y'	–	Y	Y	N	N	N
C	SUPPLIER OF MASTER-FILE-RECORD = 'V'	–	–	–	Y	Y	N
C	LEAD-TIME OF MASTER-FILE-RECORD NOT > REQUIRED OF TRANSACTION-RECORD	–	Y	N	Y	N	–
A	MOVE "BEARING" TO ORDER-ITEM MOVE QUANT OF BEARING OF TRANSACTION TO ORDER-QUANT.	X	X	X	X	X	X
A	PERFORM FILL-ORDER.	X	–	–	–	–	–
A	PERFORM ENGINEERING-ADVICE. PERFORM RFQ. PERFORM MODEL-SHOP-ORDER.	–	–	–	–	X	X
A	PERFORM BACK-ORDER ADD QUANT OF BEARING OF TRANSACTION TO BACK-INVENTORY OF MASTER-FILE-RECORD. PERFORM WORK-ORDER.	–	X	X	–	–	–
A	PERFORM BACK-ORDER ADD QUANT OF BEARING OF TRANSACTION TO BACK-INVENTORY OF MASTER-FILE-RECORD. PERFORM PURCHASE ORDER.	–	–	–	X	–	–
A	MOVE "RUSH, TOP PRIORITY" TO ADVISE OF ORDER-RECORD.	–	–	X	–	X	X

Figure 10-3 Expansion of the macros in the Figure 10-2 decision table.

beyond the scope of this book, but the general approach used is somewhat similar to the macro definition facility in many assembly languages.

Notice that the preprocessed decision table fits the stated definition for a macro; the decision table *is* a macro in a sense. Moreover, macros can often be used within the stub and rule entries of decision tables. In fact, it is quite common to use shorthand types of macros in programming decision tables and several vendors of decision table preprocessors offer combined decision table-shorthand packages. Figure 10-2 shows a decision table containing macros in both the conditions and actions sections. These COBOL macros are compact ways of writing larger, more detailed, groups of COBOL statements which occur often in an applications area. COBOL macros are not currently widely used and their format is far from standardized. In this example, the common *assembly language macro* approach is used by putting variable parameters in parenthetical clauses following the macro *verb*. Figure 10-3 shows the same table after the macros have been expanded. The general definition of these macros and the expansion process is not given here. [However, portions of the expansion in Figure 10-3 are due to a first and second parameter.] One macro may effect the expansion of another. Such effects are illustrated in the expansions of ORDER where BEARING was set by ITEM-ORDER. Macros may be simple shorthand forms (this form is not illustrated here). However, the parameter insertions in GET-FROM are typical of multi-unit shorthand substitutions), direct insertion of parameters (illustrated by SUPPLY-OK), or change of form (illustrated by QUALIFY). These macros are meant to be close to the terminology that might be used by the system analyst.

INTERACTIVE PROGRAMMING AND DECISION TABLES

The term *Interactive programming* means that the programmer communicates directly with the computer through a console, usually a typewriter or cathode ray tube (CRT) type of device. Commercial COBOL programmers are now beginning to use interactive programming. FORTRAN and other programmers have been using it for some time.

This environment is similar to the "good old days" of programming when debugging consisted of sitting at the console and displaying core locations in mid-run, altering the stored program, and the dozens of other tricks known to console debuggers. The breed of console debuggers faded because of the increasing cost of machine time; no one could afford the high percentage of idle time associated with console debugging.

But the advent in recent years of multi-programming has changed that picture. Interactive programming as it is practiced today is conducted at the COBOL symbolic, rather than at the machine, level and doesn't permit many of the old-timer's techniques. But it does permit, typically, changing the symbolic (source) program, and inspecting results via the console, saving considerable computer turnaround time. Many systems also permit inspection of the program state in mid-execution along with other useful facilities.

Decision tables can be extremely useful in the interactive environment. They are powerful macros representing a great deal of logic in a very compact form. Their automatic logical integrity (completeness, lack of redundancy and contradiction) helps both programmers and system analysts control their logical structures. The ease with which decision tables can be changed is welcomed by the interactive programmer.

The use of macros, such as shorthand,

within decision tables has the obvious added advantage of cutting down considerably the key-strokes required on the keyboard. Thus properly designed shorthand can help directly reduce errors while increasing readability.

The documentation value of the decision table is of particular value to the on-line interactive programmer. With a relatively slow (typewriter) or limited space (CRT) display device, the compact table format delivers a lot of information at a minimum of cost.

The format of input of the decision table on the console is obviously not exactly the same as the card-oriented formats used by many preprocessors. Alignment of rules vertically has been altered by several different implementations. The authors feel that it is still too early to suggest a conventional decision-table format for use in the interactive environment.

Part IV

USES AND DEVELOPMENT OF DECISION TABLES

SECTION 11

Systems Analysis

All too often, those who are learning the techniques of decision tables tend to visualize an individual struggling with a complex set of conditions and actions and using a decision table just to solve his problem. While this individual use is both valid and frequent, we must not lose sight of the fact that decision tables were developed primarily as a vehicle for man-to-man communication. This area of communication among humans is where decision tables clearly show their value and their superiority over flowcharting, narrative descriptions, and other communication media.

This section describes the process of analyzing a commercial system—a highly communications-oriented procedure—in order to underline the value of decision tables as a tool for creating mutual understanding of a complex problem among people.

We define *system analysis* as the process through which the analyst gains an understanding of an existing situation with the objective of applying that understanding to the development of a new system to handle that situation. The development of the new system is called *system design*.

Clearly, the kind of situation studied determines, to a large extent, the relative duration and intensity of these two phases. Contrast the development of a software system—a compiler, for example—with that of a commercial application such as a payroll system. In the case of the compiler, analysis is rapid and fairly simple and consists mainly of learning the source language and the machine language. The design phase, however, is where the difficult work is done. Moreover, the entire cycle of analysis and design of the compiler is characterized by its individuality; that is, the analyst requires little contact with others, since he himself originates both questions and their ultimate answers. On the other hand, the analysis of the payroll system is likely to be long and painstaking since it involves, like most commercial projects, a great deal of information exchange between the client (the payroll department) and the analyst. In contrast to the software designer who initially possesses most of the knowledge necessary to his task, the commercial analyst has little or no knowledge of his problem area. He must collect written information, interview client personnel, and obtain approvals until he is satisfied that he knows the content of the problem he is studying.

When he and his client finally agree that the existing system is represented by the documentation they have collected and developed, the design phase usually proceeds rapidly and consists of assigning functions to computer runs and allocating information to files.

ANALYSIS

The initial steps taken by the commercial analyst are the same regardless of the technique he is going to use. These steps consist of:

1. Identifying the problem area to be studied (e.g., payroll).

2. Factoring the problem area into modules—components that are relatively self-contained so that they can be treated independently for purposes of analysis (e.g., vacation accrual, FICA, state tax).

3. Collection of documentation, especially forms containing data that is relevant to the problem.

4. Identification and naming of all data fields that impinge on the problem area.

If we assume that the existing system or problem solution (e.g., the manual payroll system) is not so clearly documented that the analyst can discover its nature for himself, the next step is that of interviewing the person most knowledgeable in each of the individual problem areas initially identified by analysis. The purpose of each interview is to determine what happens (actions) and why (conditions).

Decision tables provide an excellent framework for taking down the verbal output of the client as he provides it. Because of its structure, the decision table facilitates the capture of the rambling, nonsequential dialog that typifies an interview. As the analyst hears what he thinks are conditions ("when a journeyman works over forty hours in any seven day period . . .") or actions ("he gets time and one-half unless . . ."), he enters them in the stub portion of a table and fills in the particular decision rule (the sequence of yes's and no's) mentioned. The usual interview will cover the same conditions and actions over and over, giving different decision rules at each repetition; for example, consider the following explanation:

> When a journeyman works more than forty-hours in any seven day period, he gets time and one-half unless he works more than fifty. Time over fifty hours is double-time. Apprentices get time and one-half for overtime.

Here, the narrator is discussing two conditions—union grade and hours worked—and a set of actions concerned with payment. The decision table that might result from documenting this narrative is as shown at top of next page.

After an interview, the analyst uses the information represented in his decision table to identify areas in which he needs more information. Some questions he might ask are:

1. What is the total range of values that the variable in each condition can take (e.g., in the table at top of next page, what other union grades are there besides journeyman and apprentice)?

2. Is data available in the present sys-

PAY TABLE	1	2	3	4
JOURNEYMAN	Y	Y	N	-
HOURS WORKED > 40	Y	$	Y	N
HOURS WORKED > 50	N	Y	-	-
HOURS WORKED * RATE				X
40 * RATE	X	X	X	
(HOURS WORKED-40)*(1.5*RATE)	X	X	X	
(HOURS WORKED-50)*(.5*RATE)		X		

tem to provide answers to the conditions (e.g., are overtime hours segregated from straight time or must they be computed)?

3. Are all the conditions that pertain to the problem module shown in the table (e.g., is there an upper limit on overtime pay that was not mentioned)?

4. Is the table complete?

5. Are there redundant rules and if so do they represent errors or slips of the tongue?

6. Are there contradictory rules?

These questions provide the basis for another interview at which the table is used as the communication medium. As soon as the analyst is satisfied that all the relevant conditions are represented in the table, he may wish to expand the table so that it contains nothing but simple decision rules (no dashes). Then he and his client can verify that each simple rule is correct. (If the problem area was well documented by narratives or flowcharts of an existing system, the analyst might well have started with an expanded table.)

The preceding table would be expanded as shown below.

The decision rules 3 and 7 represent combinations of the third and fourth decision rules (from the above table) that either can't happen or shouldn't happen depending on how the input to the table is stated. Therefore these rules should be isolated in an ELSE rule.

When each rule has been verified or assigned to an *impossible* category, the table can be reduced by recombining its rules. The reduced table is shown on next page.

When the entire problem area has been reduced to a set of tables by this iterative process of interview/analysis, the analysis portion is finished. At this juncture, the existing system or problem area is represented by a fairly time and process independent set of "management" (or perhaps "people") rules contained in the tables.

Two last subjects should be covered be-

PAY TABLE	1	2	3	4	5	6	7	8
JOURNEYMAN	Y	Y	Y	Y	N	N	N	N
HOURS WORKED > 40	$	Y	N	N	$	Y	N	N
HOURS WORKED > 50	Y	N	Y	*	Y	N	Y	*
HOURS WORKED * RATE				X				X
40 * RATE	X	X			X	X		
(HOURS WORKED-40)*(1.5*RATE)	X	X			X	X		
(HOURS WORKED-50)*(.5*RATE)	X							

(Recall that the * in the entry portion is an implied No and the $ is an implied Yes. The * in the stub is the COBOL multiplication operator.)

PAY TABLE	1	2	3	4	ELSE
JOURNEYMAN	Y	Y	–	N	–
HOURS WORKED > 40	$	Y	N	Y	–
HOURS WORKED > 50	Y	N	*	–	–
HOURS WORKED * RATE			X		
40 * RATE	X	X		X	
(HOURS WORKED-40)*(1.5*RATE)	X	X		X	
(HOURS WORKED-50)*(.5*RATE)	X				
ERROR					X

fore considering the design phase. First, what language should appear within the tables prepared during analysis? Because a decision table provides only a structure, any language capable of expressing conditions and actions can be used within it. Thus any language that is understood by both the analyst and his client is satisfactory. If the client understands the language in which the tables will ultimately be presented to the computer, so much the better. Most clients, however, are more comfortable using their own "jargon" and the analyst should utilize that dialect if there is any question of whether the computer-oriented language will be understood.

Second, no mention has been made of the size of the tables being developed. Are there limits to the number of conditions, actions or decision rules that should be portrayed in a single table? The preceding discussion is concerned with the process of analyzing systems which is, as we saw, largely an exercise in communication between people. For this communication to be effective, the decision tables that are its subject must be readily grasped by the communicators. Thus the general rule is that decision tables used for man-to-man communication should be no larger than either party to the communication can easily comprehend. Experience indicates that a table with more than sixteen written decision rules is difficult for most individuals to grasp. This would limit tables with simple rules to no more than four conditions. If complex rules or the ELSE rule or both are used, then the number of conditions could be increased to five or at most six without loss of the capability to grasp the meaning. Because the actions do not effect the complexity of decision tables to as great a degree as conditions, more actions can appear without clouding the issue.

DESIGN

The design phase is the first in which the computer intrudes. Throughout analysis, an existing system was explored and documented with little regard as to the way it would eventually be implemented. During design, however, the processing medium must be considered to some degree. How much of an impact it has depends on such factors as the organization of the EDP department and the relative skills and personalities of the designer and programmer (if indeed they are different individuals). If the designer merely indicates the content of the system's files and records and the layout of reports and defines the overall sequence of events within the system, then the computer intrudes very little, If, on the other hand, he produces detailed record formats and groups events into a set of programs, then the intrusion is greater. This latter is generally the more common of the two.

Whatever the degree of intrusion, the primary objective in a design using decision tables is to separate as completely as is possible the computer oriented processes from the human oriented. More specifically, the content of the tables developed during analysis should be preserved as is. While this objective is seldom, if ever, completely attainable, it is worthwhile to come as close as one can. The reason is that the client should be able to understand and verify his tables at any point in the implementation cycle. If the tables that contain his rules are extended to contain computer-oriented conditions or actions, the client will be less able, or perhaps completely unable, to make this verification.

Therefore the design phase occurs in three stages:

1. The decision tables resulting from analysis are placed in the correct sequence, then combined or split and recombined to simplify them and correct for overlap caused by the original factoring of the problem area.

2. Additional decision tables and "in-line" routines are specified to implement the analysis tables on the computer.

3. The specification of linkage between the analysis tables and this support code completes the design process.

PROGRAMMING

The separation of system-oriented and computer-oriented processes should be continued throughout the programming phase. This is especially critical when the language used within a decision table must be translated from the client's language to a computer language. Two practices that help to preserve the clarity of the system tables are the following:

1. Use of condition-name tests wherever possible to enhance the readability of the table; for example, use the condition

IF JOURNEYMAN

[where JOURNEYMAN is a condition-name (level 88)] rather than the comparable relation test

IF UNION-GRADE EQUAL '2'.

2. Convert system-oriented actions to COBOL PERFORM statements of meaningful procedure-names whose procedures do computer-oriented processing; for example, the system-oriented action

open customer account

can be restated as

PERFORM OPEN-CUSTOMER-ACCOUNT

where OPEN-CUSTOMER-ACCOUNT is a separate in-line routine or decision table (or both) that performs the computer-oriented tasks necessary to create a new customer master record.

We assume that a decision table preprocessor will be used to convert the combined tables and in-line code to compiler-acceptable code. Where a preprocessor is not available, the programming step must include the hand conversion of tables to code. It is essential that when hand conversion methods are employed, the separation of conditions and actions in each table is maintained in the converted code in order to facilitate later modification of the table and its code without major reconversion.

TESTING

To the extent that the system-oriented and computer-oriented processes have been kept separate, testing is a two level process. The client, assisted by the analyst, prepares test data to test the systems-oriented tables, while the programmer prepares data to test the computer-oriented processes. In both cases, data is prepared such that each decision rule in each table (including ELSE) is traversed by at least one transaction.

During the testing phase, the client is primarily interested in verifying that the computer system solves his problem—does it follow the rules that he and the analyst have coded in the system-oriented tables? The programmer is mainly concerned with verifying that the programs in the system are sound in terms of computer processing —does each program do what he intended it to do without breaking down? The separation of system-oriented and computer-oriented processes permits both questions to be answered by the parties best qualified to ask them. Without such separation, the programmer is often the only one able to adequately test the system. Thus it is not surprising that many systems run correctly according to the programmer's criteria but fail to solve the client's problem.

MAINTENANCE

Decision tables provide excellent documentation of a system, not only because of their inherent clarity of expression, but also, if a preprocessor is used, because they exactly represent the computer program. This exactness of representation, furthermore, does not diminish with the passage of time, as it does with flow-charts and narratives.

Where the division of system and computer-oriented processes has been maintained throughout implementation, decision tables provide an added maintenance bonus —the client can specify changes to the system almost directly to the computer. Because the existing system is represented by a set of decision tables that he understands, the client can easily evaluate the impact of system changes and specify them very precisely. He can also provide changes or additions to the system test data to check-out the revised system. This ability of the client to specify system changes in terms of actions to be taken rather than in desired results simplifies the maintenance process while preserving the client's understanding of his system.

Logical Organization

As in programming applications, decision tables are used in system analysis for handling the logical aspects of the situation. It is the relationship of cause (Conditions) to effect (Actions) that forms the logical structure of a system.

Causes, as for example a man's pay rate and his hours worked during the week, lead to an effect, which in this case might be calculation of his weekly gross pay. The combination of these two causes that lead to the effect constitutes a particular logical relationship, even though the specific calculation required to get from rate and hours to gross pay may vary from analyst to analyst. This might be likened to a high-speed expressway. A road may go from Chicago to New York, but whether you travel in a sports car, limousine, or ten-ton truck is immaterial from the logical point of view.

In data processing systems, the logical structure is the route between the source of data and the final production of information. How these routes are ultimately traveled is a matter of processing method and not of the logical structure with which the system analyst is concerned.

Interfaces

An interface might be defined as "where things meet." From the system analyst's point of view, an interface is where data meet and enter the system and where information leaves the system as output. Data is used here to mean the raw input, unprocessed by the particular system into which they are being entered; information is defined here as that which the system has extracted from the data.

Breaking down a system into parts means adding new interfaces between these parts. Combining separate systems means eliminating common interfaces. Redefining a system means finding entirely new interfaces.

ROLE OF THE SYSTEM DESIGN TABLE

Decision tables for system design purposes are intended primarily to help the analyst study the structure of his system. Using the system design table, the input data are associated with the output information in the system to enable the analyst to recognize the natural divisions of the system and logical similarities between systems. Helping the programmer develop programs to handle the details of that system is a secondary benefit of these tables. Thus the language used is nearer to a natural rather than a programming language.

The system design table can be analyzed in various ways, and these will be discussed here largely by example. Objectives of such analysis include determining:

1. Which input data are really required and which are redundant.
2. How a system might be divided.
3. How systems may be combined.

Admittedly, this whole area is far more of an art than a science—few algorithmic guidelines exist. However, in the list above, (1) is somewhat similar to redundancy and contradiction checking for decision tables, (2) to the splitting of tables, and (3) to the combination of programming types of decision tables—all topics that are discussed elsewhere in this book. Interfaces are represented in a system analyst's decision table as conditions for input data and as actions

STUBS...		RULES...			
C	Pay-Rate	N	N	Y	Y
C	Hours-Worked	N	Y	N	Y
A	Gross-Pay				X
A	Error	X	X	X	

Figure 11-1 Decision table for description of gross pay problem.

for output information. As a description of the previously mentioned gross pay problem, a system analyst's table might be as shown in Figure 11-1.

Here, two input data interfaces (Pay Rate and Hours Worked) are expressed as conditions, while the information interfaces (Gross Pay and Error) are expressed as actions.

Every combination of the conditions is included within the rules. Notice that Y and N are used (other condition rule entries will be discussed later) to express presence or absence of the condition. Only one of these combinations leads to calculation of a gross pay figure. How that figure is calculated is unimportant at present.

The other three rules define errors and an action is prescribed for this. The presence of a logical error in the input data is always a valid item of information, and thus an action in the system design table and an output of a system.

Setting Up The System Design Table

The system design table may be set up as a binary matrix table where every combination of truth and falsity of every condition is included. Where many types of data are involved such tables would become very large, so that other entries such as don't-cares and extended may be desirable. Dependencies may also exist. Where these dependencies are apparent it may be advantageous to use * and $ to improve the readability of the

Input Data	Output Information
Time-cards	Pay-checks
Gross-adjustments	Labor-Distribution
Pay-rate	Payroll-Register
YTD-earnings	YTD-Earnings
Tax-Table (Federal)	Bond-Register
Tax-Table (State)	Stock-Register
Tax-Table (City)	Garnishee-Checks
Deduction-Master	
Insurance	
Garnishees	

Figure 11-2a Input/output description for a simple payroll system.

table. Our examples will utilize only Y, N, and – (don't-care).

Figure 11-2a is an expansion of our earlier gross pay example into a simple payroll system for purposes of illustration.

Reduction of Table Size

There are ten data inputs in this example. To form a binary table—one that includes every possible combination of these ten conditions—would require 2^{10} or 1,024 rules.

A table of 1,024 rules, as required for the rules to be generated from Figure 11-2a,

Input Data	Output Information
Gross-affected by	(Time-cards, Gross-adjustments, Pay-rate)
Year-to-Date by	(YTD-earnings)
Taxes by	(The three Tax Tables)
Deductions by	(Deduction-master, Insurance, Garnishees)

Figure 11-2b Simple payroll system with conditions grouped.

would be unwieldy and difficult to manage. Two techniques are available for reducing this table—and any large table—to manageable size. The first is by use of the don't-care condition, and the second is by grouping of conditions. We shall utilize both methods, grouping conditions first and then introducing don't-cares by the process of rule combination.

Grouping of conditions requires some knowledge of how these conditions will be used. In the payroll example of Figure 11-2a the conditions are grouped as shown in Figure 11-2b.

These four conditions groups may be used to form a table of only $2^4 = 16$ rules, as shown in Figure 11-3.

STUBS...	RULES...															
	1	2	3	4	5	6	7	8	9	10	11	12	13	14	15	16
Gross-affecting	Y	Y	Y	Y	Y	Y	Y	Y	N	N	N	N	N	N	N	N
Year-to-date	Y	Y	Y	Y	N	N	N	N	Y	Y	Y	Y	N	N	N	N
Taxes	Y	Y	N	N	Y	Y	N	N	Y	Y	N	N	Y	Y	N	N
Deductions	Y	N	Y	N	Y	N	Y	N	Y	N	Y	N	Y	N	Y	N
Pay-checks	X															
Labor-distribution	X	X	X	X	X	X	X	X								
Payroll-register	X															
YTD-earnings	X	X														
Bond-register	X		X		X		X		X		X		X		X	
Stock-register	X		X		X		X		X		X		X		X	
Garnishee-checks	X		X		X		X		X		X		X		X	

Figure 11-3 Decision table for simple payroll system with grouped conditions.

ROLE OF THE SYSTEM DESIGN TABLE

The "X" is placed in each appropriate entry when an affirmative answer applies to the question "which actions result from this particular combination of conditions?"

Analyzing the System Design Table

The system design table may be used to analyze the order in which the input data is required within the system and the order in which the output information is to be produced. One general approach is to repeat the table's conditions, each with its group of like action row(s).

This is illustrated in the upper portions of Figures 11-4, 5, 6, and 7, which are derived from Figure 11-3 by repeating the conditions of the table shown in Figure 11-3, but each with its own unique series of actions. Those rules that have no actions in any particular table are assumed directed to the Else rule and, in effect, dropped from that table.

In each of these new tables, redundancy exists among the set of rules, since each has the same action. These redundant rules

STUBS...	RULES...							
	1	3	5	7	9	11	13	15
Gross-affecting	Y	Y	Y	Y	N	N	N	N
Year-to-date	Y	Y	N	N	Y	Y	N	N
Taxes	Y	N	Y	N	Y	N	Y	N
Deductions	Y	Y	Y	Y	Y	Y	Y	Y
Bond-register	X	X	X	X	X	X	X	X
Stock-register	X	X	X	X	X	X	X	X
Garnishee-checks	X	X	X	X	X	X	X	X

STUBS...	RULES...
Gross-affecting	–
Year-to-date	–
Taxes	–
Deductions	Y
Bond-register	X
Stock-register	X
Garnishee-checks	X

Figure 11-5 Transformation of rules 1, 3, 5, 7, 9, 11, 13, and 15 of figure 11-3.

STUBS...	RULES...							
	1	2	3	4	5	6	7	8
Gross-affecting	Y	Y	Y	Y	Y	Y	Y	Y
Year-to-date	Y	Y	Y	Y	N	N	N	N
Taxes	Y	Y	N	N	Y	Y	N	N
Deductions	Y	N	Y	N	Y	N	Y	N
Labor-distribution	X	X	X	X	X	X	X	X

STUBS...	RULES...
Gross-affecting	Y
Year-to-date	–
Taxes	–
Deductions	–
Labor-distribution	X

Figure 11-4 Transformation of rules 1-8 of figure 11-3.

STUBS...	RULES...	
	1	2
Gross-affecting	Y	Y
Year-to-date	Y	Y
Taxes	Y	Y
Deductions	Y	N
YTD-earnings	X	X

STUBS...	RULES...
Gross-affecting	Y
Year-to-date	Y
Taxes	Y
Deductions	–
YTD-earnings	X

Figure 11-6 Transformation of Rules 1 and 2 of figure 11-3.

82 SYSTEMS ANALYSIS

STUBS...	RULES...
	1
Gross-affecting	Y
Year-to-date	Y
Taxes	Y
Deductions	Y
Pay-checks	X
Payroll-register	X

STUBS...	RULES...
Gross-affecting	Y
Year-to-date	Y
Taxes	Y
Deductions	Y
Pay-checks	X
Payroll-register	X

Figure 11-7. Transformation of rule 1 of figure 11-3.

may be combined. Combination procedures are discussed elsewhere in the book in detail, but here it suffices to say that if two rules have the same actions and differ in only one condition entry—a Y and an N—then the two rules may be combined. The differing entry becomes a don't-care (−).

The final result of full rule combination of the four new tables is shown in the lower portions of Figures 11-4, 5, 6, and 7. It should be noted that in our example all rules were able to be combined; this will not always occur.

By combining rules in each of these tables, the order in which input data is needed and the order in which output information is produced within the system may be deduced. This demand for data (order in which input is needed) can be considered a major criterion in system design.

Application of the Analysis to Program Flow

The program flow for the simple payroll system used as our example can be shown based upon the original decision table of Figure 11-3. The resultant block diagram (Figure 11-8), however, is gross and unwieldy and serves little more than to show the basic data-to-information relationship.

The advantage of analyzing the system using the system design decision tables becomes apparent in Figure 11-9. Based on the results of analysis presented in the lower portions of Figures 11-4, 5, 6, and 7, Figure 11-9 reveals that the payroll system can also be divided into as many as four parts. The system is now "open" for definitive analysis.

Before leaving the payroll example, it should be pointed out that it has been highly simplified in order to illustrate the technique involved. Mainly, deductions have been assumed completely independent of

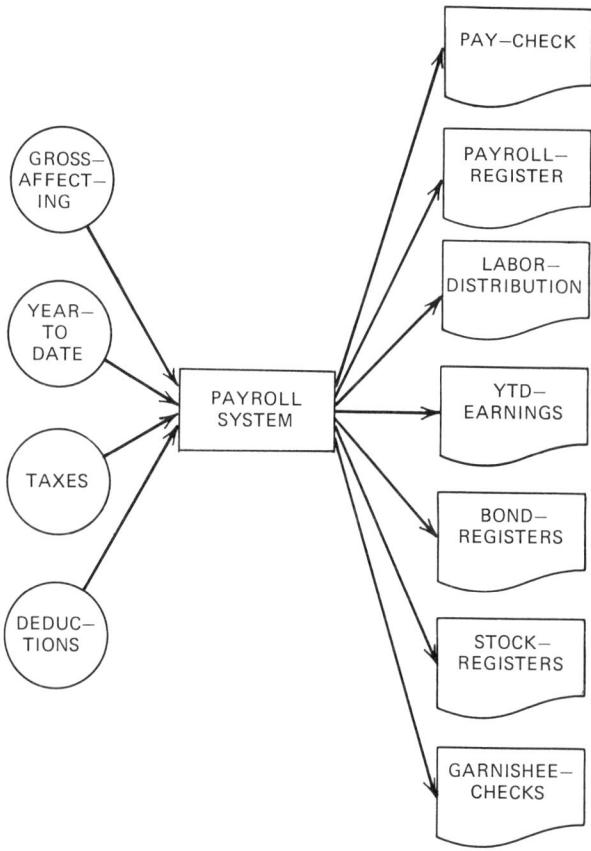

Figure 11-8 Block diagram of payroll system based on Figure 11-3.

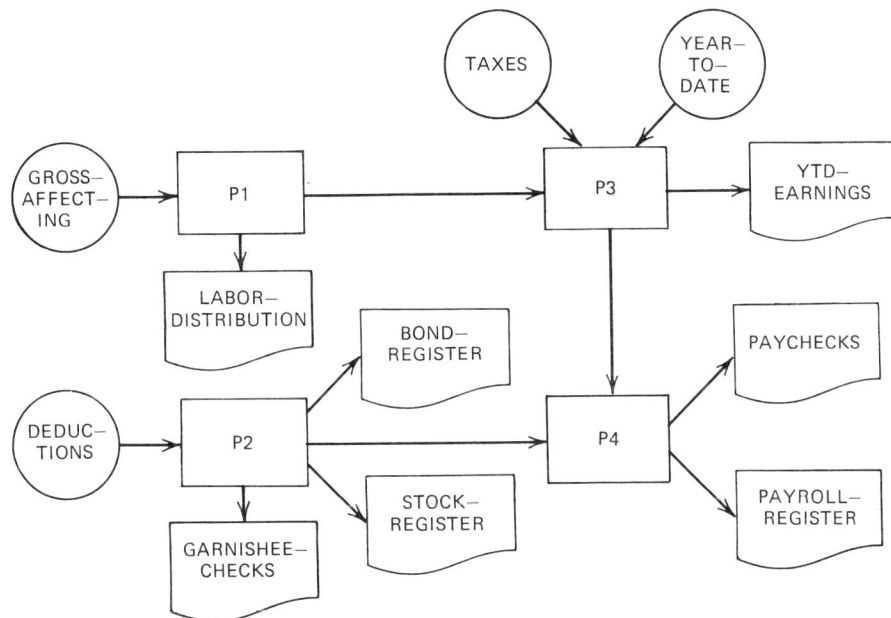

Figure 11-9 Block diagram of payroll system based on Figures 11-4, 5, 6, and 7.

gross, a situation that is seldom true in real systems. However, this does not invalidate the technique; the groupings assumed in Figure 11-2b could be changed to match the correct assumptions.

Use of the Don't-Care and Other Condition Types

Don't-care entries (−) were introduced during the preceding payroll example only in the final step of rule combinations (Figures 11-4, 5, 6, and 7). In more practical cases, the approach just outlined will result in large unwieldy tables. Their size may be reduced initially by determining what relationships the various data inputs have to one another. Primarily, certain ones will have an effect that obviously doesn't depend on the others. An example is the "Deductions" group of entries in Figure 11-3.

Given the restrictions stated in that example, deductions may be used to produce the three deduction registers regardless of the other three inputs, leading us directly to the table shown in the lower portion of Figure 11-5.

The same is true of the Labor-distributions, assumed here to be in terms of man-hours only and thus obtainable from only the gross-affecting input. This leads us to the bottom table in Figure 11-4, containing don't-cares.

Only less obvious cases may not need the full analysis, such as that in Figure 11-6. Once the simplified one-rule tables are obtained for each action, the relationship of one to another, in terms of data demand, becomes visible.

SUMMARY

Decision tables work to their full capacity when used to facilitate man-to-man communication for complex decision rules. Tables should be used from the inception of analysis through system documentation and maintenance. The degree to which they benefit a system development project depends on the following:

1. How careful the analyst is in insuring that his client completely understands the tables developed during analysis.

2. How closely the analysis tables represent the system being studied.

3. How carefully the analyst has analyzed the tables for proper structuring of the application.

4. How well the separation of system-oriented processes from computer-oriented processes is maintained through the design and programming stages.

SECTION 12

Decision Tables in Computer Programs

As noted earlier, a major roadblock to the expanded use of decision tables has been the absence of decision table preprocessors to automatically convert decision tables to computer coding. Despite the many advantages that decision tables offer, computer programmer/analysts were reluctant to use decision tables if they had to perform the additional chore of manually converting the decision table to coding (or its equivalent, a flowchart). Fortunately, within the past two years, a sizable number of good decision table preprocessors have been developed and marketed so that a significant increase in the use of decision tables can be expected.

Rather than use one of the commercial preprocessors for illustrative purposes, we use a fictitious preprocessor to be known as *DECITAB*. It will incorporate features from some of the current better known preprocessors, as well as features that are desirable but not necessarily implemented in existing preprocessors. No formatting rules are assumed for *DECITAB*.

The particular preprocessor to be used for converting the table into programming language code determines how this basic structure of stubs and rule entries should be arranged. Detailed format instructions are best obtained from the organization whose preprocessor is being used.

COBOL AND DECITAB

Computer programs, including the decision tables contained within them, can be written in any one of a myriad of languages, for example, COBOL, FORTRAN, PL/1, machine language, or assembly language. The decision tables of a COBOL (or other language) program must be converted into that programming code by means of a preprocessor computer program.

For convenience and consistency when dealing in the text with computer program decision tables, we refer to decision tables that contain COBOL. We assume that the reader is already familiar with COBOL or will achieve the requisite familiarity by referring to Appendix II, *An Introduction to COBOL*.

The following are three approaches to processing decision tables that are part of COBOL programs.

1. Have the COBOL compiler accept them as extensions of the language; decision tables are currently not part of ANS COBOL.

2. Arrange for the decision tables to be accepted as data to be interpreted during execution of the computer program.

3. Convert the decision tables to COBOL source statements prior to compilation by the COBOL compiler.

DECITAB uses the third approach.

We will be discussing two sets of decision table components. The first is the specific set of components for our hypothetical preprocessor DECITAB; the second is an abstracted set of components available for decision tables that can be used for problem definition and analysis and are not part of a computer program. Format details for DECITAB are not provided, since such detail would lend no particular utility to our discussion.

DECITAB Components

DECITAB tables are entered on the following set of records:

1. DECITAB control record, which contains the table name and other control items.

2. Entries that specify the rule numbers and delineate the rule entry boundaries.

3. Initialization statements containing the series of activities that may be caused to occur prior to each of the rules in the tables (their execution is optional on each entry to the table).

4. Condition row records, each containing a stub and one or more rule entries.

5. Action row records containing fields similar to those of the condition rows.

6. The frequency distribution record, which describes the expected relative frequency with which each rule will be satisfied by transactions.

7. The table delimiter assumed here to be denoted by TABEND, an abbreviation for "Table end."

Actual preprocessors of the DECITAB type include, at the time of this writing, DETRAN, DETOC, DETAP, DETAB-70, TABTRAN, SMP, DECISUS, and others. Many features of each of these are reflected in DECITAB, but this hypothetical preprocessor contains some features not found, at the time of writing, in any of the above.

In the nonfictitious preprocessors, the "record" used above is usually a unit-record card or, in on-line console applications, an input stream. Because of their limited size, cards (the usual input medium) present certain technical considerations such as rule and stub continuation. These considerations are handled differently by real preprocessors, so that any convention adopted here would be of little generality. Instead we take the "input stream" approach and assume an unlimited record.

Decision tables enable the programmer or analyst to concentrate on the problem logic rather than the details of evaluating the logic and can be used for describing those program segments that deal with logical structures. Decision tables can be used

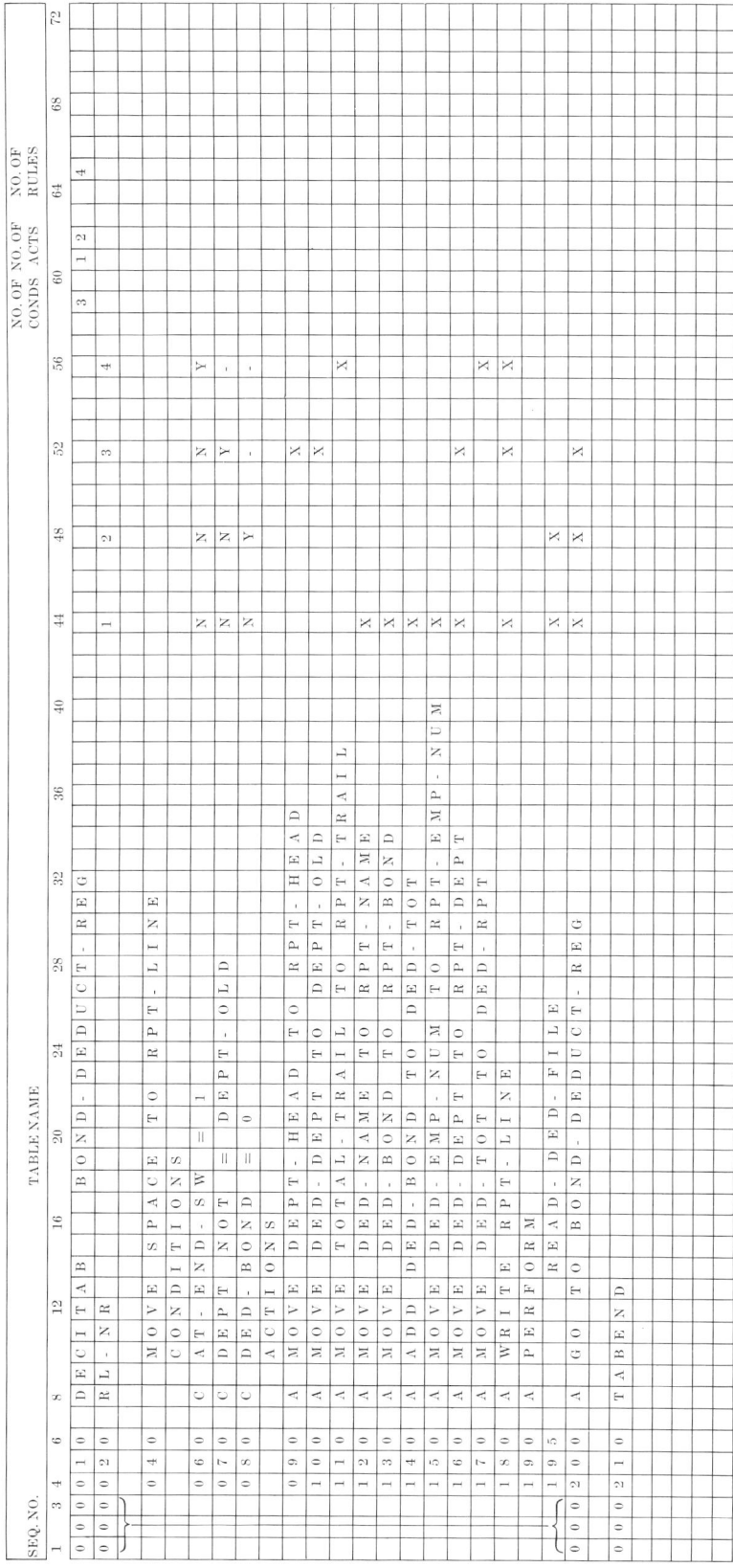

Figure 12-1 Sample DECITAB decision table.

for systems analysis, design and program development. Where they are used for computer program development, it is extremely useful to have available a preprocessor to convert the decision tables into programming language code.

Let's look at a typical DECITAB table to see how it works. This table (Figure 12-1) represents the logical problem of producing a simple report that contains a bond deduction register by individual by department.

The table in Figure 12-1 shows a DECITAB table, containing COBOL language in the stubs, written on a coding sheet. Columns 8 and 9 identify the type of row; for example, "C" followed by a blank signifies a condition. Note that a row (initialization, condition, or action) can occupy more than one physical line on the form; for example, the action row described on lines 190 and 195.

This table contains four rules, with the four one-digit entries in the RL-NR cards specifying the beginning columns in which rule entries are located. Those rows containing a "C" in Column 8 are conditions. Their stubs contain the condition to be satisfied as specified in the rule entries in order for the series of actions (those rows with "A" in Column 8) to be executed for the selected rule.

The table contains one initialization statement (statements that precede the conditions) that moves spaces to the entire print record. The table can be entered from other points or can loop through itself (via Rules 1, 2, and 3). It prints register lines for each of the deduction records, showing bond deductions that it reads from the deduction file. It also prints department headings, assumed to be the major sequence of the department file.

The four rules are related to the following four situations that might be encountered:

1. A record that has a bond deduction has been read.
2. A record that has no bond deduction has been read.
3. A new department has been encountered; this requires a department heading to be printed; or
4. The end of the file has been reached, in which case a report trailer line is to be printed along with the total deductions.

The four rules of the table correspond to these four situations and may be stated, in COBOL terms, as shown in the next two pages. Several more comments about this table are in order.

First, notice that not more than one rule can be true at any one time. Between any two rules, at least one condition will have a Y for one rule's entry and an N for the other's entry.

Next, notice that this table is *complete*, that is, any possible combination of the three conditions will lead to one of the four rules. Why this is so is discussed in the section on completeness of tables. If this table had not been complete, an ELSE rule would have been required.

THE HOST PROGRAM FOR A SET OF DECISION TABLES

A table such as the one just presented is not a program; rather, by itself or along with other decision tables, it is a logical structure within a program that acts as the tables' host. In this book, we assume that tables can be included in conventional COBOL programs as an extension to the COBOL language, and that the combination is processed through an appropriate preprocessor before compilation. When this is done, a table is equivalent to a COBOL Procedure Division SECTION. It may be performed, referred to by a GO TO, or simply

COBOL Equivalent of Rule 1		Notes

IF DED-BOND NOT = 0
   AND
   DEPT = DEPT-OLD
   AND
   AT-END-SW NOT = 1

| C | The NOT is due to the N in
| O | the rule entry.
| N | The NOT in the stub is negated
| D | by the NOT in the entry.
| I | This record does not start
| T | a new Department. It is
| I | assumed that a programmed switch,
| O | called AT-END-SW, is set when-
| N | ever an end-of-file condition
| S | is detected.

THEN
MOVE DED-NAME TO RPT-NAME,
MOVE DED-BOND TO RPT-BOND,
ADD DED-BOND TO DED-TOT,
MOVE DED-EMP-NUM TO RPT-EMP-NUM,
MOVE DED-DEPT TO RPT-DEPT,
WRITE RPT-LINE,
PERFORM READ-DED-FILE,

GO TO BOND-DEDUCT-REG.

| A |
| C | Print-line is built and a
| T | total (DED-TOT) is developed.
| I |
| O |
| N | Report line is written
| S | New record is read by a PERFORMed
routine. If the end of the file
has been reached, an AT-END-SW
switch is set in that routine.
The table is re-entered

COBOL Equivalent of Rule 2		Notes

IF DED-BOND = 0
   AND
   DEPT = DEPT-OLD
   AND
   AT-END-SW NOT = 1

| C | Same as Rule 01 but there
| O | is no bond deduction (DED-
| N | BOND = 0) for this record.
| D |
| I |
| T |
| I |
| O |
| N |
| S |

THEN
PERFORM READ-DED-FILE,
GO TO BOND-DEDUCT-REG.

| A | A new record is read in a
| C | PERFORMed routine which also
| T | sets AT-END-SW switch if needed
| I | and the table is re-entered
| O | without having printed a
| N | register line.
| S |

DECISION TABLES IN COMPUTER PROGRAMS

COBOL Equivalent of Rule 3

IF DEPT NOT = DEPT-OLD
   AND
   AT-END-SW NOT = 1

CONDITIONS	Notes
	Department number has changed on this record. Note that presence of a bond deduction is immaterial.

THEN

MOVE DED-DEPT TO DEPT-OLD,
MOVE DEPT-HEAD TO RPT-HEAD,
MOVE DED-DEPT TO RPT-DEPT,
WRITE RPT-LINE,

GO TO BOND-DEDUCT-REG.

ACTIONS	Notes
	New Department number is recorded.
	A special header is set up. Special department header printed.
	Table is re-entered. (Note that no new record has been read, so the previous record is still available for evaluation.)

COBOL Equivalent of Rule 4

IF AT-END-SW = 1

CONDITIONS	Notes
	The last reading of the deduction file encountered an end-of-file indicator.

THEN
MOVE TOTAL-TRAIL TO RPT-TRAIL,
MOVE DED-TOT TO DED-RPT,
WRITE RPT-LINE.

ACTIONS	Notes
	The total line is set up and printed.
	(Note that the table is *not* re-entered since its job is now done.)

"fallen into" from preceding code. In turn, the table itself may GO TO or PERFORM any COBOL procedure-names or table-names in the program, or it may "fall through" into the code following the table.

Decision tables are often—but not exclusively—viewed as closed subroutines that are grouped at the end of the program and referred to by PERFORM statements.

Taking this approach, the Procedure Division of the host program for the table developed earlier might be:

```
MAIN-PROGRAM SECTION.
MAIN-PROGRAM-PARAGRAPH.
    OPEN INPUT DED-FILE,
        OUTPUT RPT-FILE
    PERFORM READ-DED-FILE.
    PERFORM BOND-DEDUCT-REG.
    CLOSE DED-FILE, RPT-FILE.
    STOP RUN.
READ-DED-FILE.
    READ DED-FILE
        AT END MOVE 1
        TO AT-END-SW.
```

The decision table previously described, BOND-DEDUCT-REG, would appear after the above coding. In actual practice, the extra PERFORM READ-DED-FILE could have been incorporated into the table, and other factors such as page ejects and more sophisticated accounting techniques would have been used.

SECTION 13

Writing the Decision Table

This section covers one of the most important topics in this book, yet it is one of the shorter sections. The reason for this brevity is that the entire book is really on the subject of writing the individual decision table.

Many of the factors to be taken into consideration when the systems analyst or programmer takes pencil in hand and starts to write a system containing decision tables are directly related to what is to be done with the tables. Evaluation of the logic or decomposition for systems analysis purposes by the writer permits a freer approach to decision tables than does a system meant to be run through a specific preprocessor and language compiler.

The individual manuals or guides to the appropriate preprocessors and compilers must be consulted for detailed information. Here we offer a general checklist of factors that may require attention; enclosed in parentheses are the references to other sections in which the particular factor is discussed.

CHECKLIST OF CONSIDERATIONS FOR DECISION TABLE WRITERS

A. **Overall Systems Analysis**

1. Division of system into manageable parts (*Systems Analysis*)
2. Combination of parts of the system (*Systems Design*)
3. Inter-relationship of parts (*Structures of Tables*)
4. Determination of the decomposition method—preprocessor or manual—and language to be used (*Languages*)

B. **Tables Intended for Decomposition**
 1. Coding conventions and formats (see the manual or guide for the preprocessor to be used)
 2. Features usable (see the manual or guide for the preprocessor and compiler to be used)
    (a) Conditions (*Conditions.*)
       (i) Are normal extended entries permitted?
       (ii) Are nonoverlapping bounded ranges permitted in extended entries?
       (iii) Are overlapping bounded ranges permitted?
       (iv) Are dependencies between condition rows expressible with *, $ entries?
       (v) Are any special condition types, such as the colon (:) entry, permitted?
    (b) Action (*Actions and Rules*)
       (i) Are action sequence numbers permitted?
       (ii) Are extended entry action rows permitted?
       (iii) Are there any special conventions concerning re-entering a table or entering other tables?
       (iv) What is the mechanism for performing other tables?
       (v) Can initialization be used and can it be bypassed or controlled?
       (vi) Do segmentation or interprogram communication conventions of the language compiler have any effect on the use of CALL as an action? Are there any passed-parameter conventions?
       (vii) Are PERFORM or other action-grouping conventions permitted?
    (c) Rules (*Rules, Contradiction and Redundancy, and Completeness Checking*)
       (i) Is run time optimization by rule frequency permitted? (*Optimization*)
       (ii) Is automatic rule combination performed by the preprocessor? (*Maintenance*)
       (iii) Does the processor do complete redundancy and contradiction checking?
       (iv) Is completeness checking done and what are the conventions for missing and for un-needed ELSE-rules?
    (d) Miscellaneous Check Items
       (i) Is initialization permitted by the preprocessor? In what format? Is it bypassable or executable as an action?
       (ii) Is an entry point to table other than the table name permitted?
       (iii) What special optimization features are available?
       (iv) The table writer should check features list and sales material for any additional capabilities of the preprocessor that may prove useful to him

C. **Selecting Decision Table Features To Be Used**
 1. How many tables are needed? (*Structures of Tables*)
 2. Naming conventions to be used (check local standards or conventions for this).
 3. Isolate situations that represent rules. (*Systems Analysis*)
 4. Isolate conditions that lead to these rules. (*Conditions*)

(a) Can they best be expressed as limited or extended entry?
(b) Are there dependencies between condition rows? If so, should *, $ entries be used if available?
(c) Are ranges of values involved as conditions?

5. Actions. (*Actions*.)
   (a) Should limited or extended entry be used?
   (b) Should action sequence numbers be used?
   (c) Should action execution sequences "fall out" the end of the table or rather be terminated by GO TO or other statements?

6. Should optimization features be used? (*Optimization*)
   (a) Should action groupings be used? (*Actions*)
   (b) Are time or core optimization techniques applicable?

7. Are initialization statements required?

D. Checking the Table[1]

1. Is the table complete? (*Completeness Checking*)

[1] Because checking the table is an important aspect of debugging it is discussed in the section, *Debugging Programs Containing Decision Tables*.

(a) Was the ELSE-rule used where needed?
(b) Were all pertinent conditions included?
(c) Were all pertinent rules represented?

2. Is the table free of contradiction and redundancy? (*Contradiction and Redundancy*.)

The simple manual checking procedure described in the section on debugging should be applied. It will help determine logical as well as transcription errors. This is sometimes called *Logical Integrity Checking*.

Writing the table is the important step at which the abstract concept of a logical system becomes a manageable and usable concrete object. The mechanics of transferring the written table into computer input, including steps such as keypunching, should be thoroughly audited to ensure that a clerical error does not creep in as well as to locate those introduced during earlier stages. All too often, unfortunately, a clerical or careless type of error "makes sense" when later interpreted as part of a system.

Detecting such errors, as well as all other types of errors, introduces the subject of debugging, the topic of the next section.

SECTION 14

Debugging Programs Containing Decision Tables

Debugging can be defined as the process of converting a computer program that was designed and developed to handle a particular problem into one that actually does handle the original problem (See figure 14-0 on next page).

Decision tables enable the programmer to write far more accurate and nearly correct code than he could write using more conventional techniques, but there is still no guarantee that the result will be correct. Debugging procedures for programs containing decision tables are able to concentrate on testing and verifying the correctness of logical relationships within the table without having to deal with the logical relationships represented by coding.

As this shift of emphasis might suggest, special techniques have been found useful in the process of debugging programs containing decision tables; these are discussed in this section.

THE LANGUAGE, THE HOST PROGRAM, AND THE TABLES

A distinction must first be made between the tables and the computer language in which both the tables and the host program are written. The language is usually determined by the programmer or by some standard in effect at the using organization.

Individual decision table preprocessors usually develop,

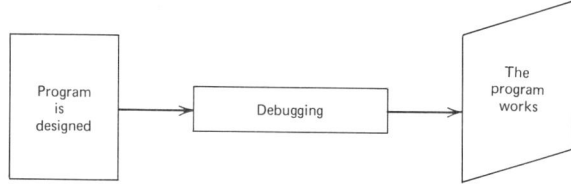

Figure 14-0 Debugging.

from the decision table, computer coding suitable for some particular language such as COBOL, FORTRAN, ALGOL, or PL/I. As a standard base for discussion, this book assumes COBOL as the programming language.

Since system design analysis tables are usually not intended for direct inclusion in a computer program, their authors have considerable leeway in the selection of the language to be used. Generally, it is not one of the normal computer programming languages at all. As for decision tables used for computer programs, the leeway in languages that is permitted is a function of the particular decision table preprocessor to be used.

We assume that the language used for the decision tables agrees with that implemented by the language translator (compiler) that will eventually be used to convert the program from source language to executable form. This aspect of decision tables, namely, checking for the proper use of language syntax and semantics, is beyond the scope of this book. Except for some comments in connection with extended entries, language is assumed to be correct in our discussion of debugging.

Decision tables must exist within a program; they do not constitute programs in themselves. We refer to the program in which one or more tables are to be included as the "host program," as illustrated in Figure 14-1.

The Procedure Division of a host pro-

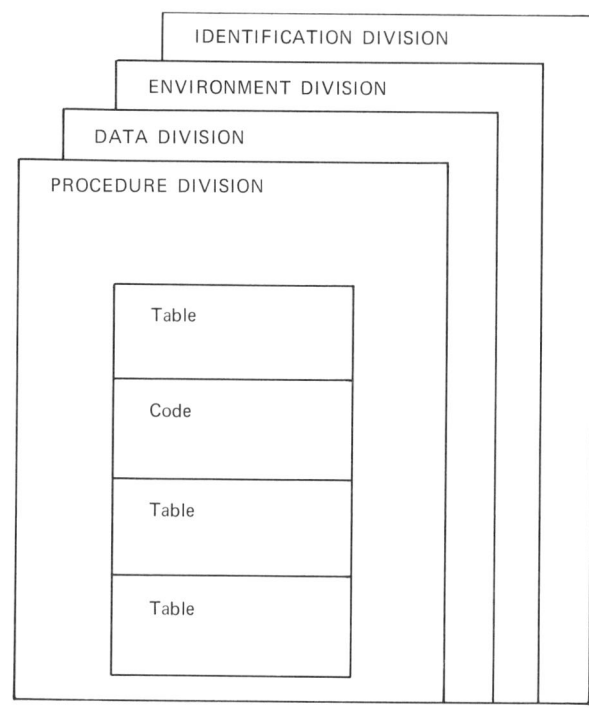

Figure 14-1 A typical host program containing a mixture of tables and conventional coding.

gram may utilize decision tables for only one small portion or for the entire logical control, leaving manipulative (or housekeeping) functions to conventional coding. It may even be expedient to have the entire procedure consist of decision tables. In any case, any one table may be thought of as separate from the rest of the program.

The debugging procedures to be discussed in the following sections cover: (a) manual checking of each individual table; (b) manual checking of the relationship of each table to the host program; (c) preprocessing of the tables; (d) compilation of the host program containing the tables; (e) tracing the execution of the host program; and (f) test data generation.

A familiarity with conventional debugging procedures is assumed. The emphasis here is on how the presence of tables alters and simplifies these procedures.

Manual Checking of Table Components

Each table, after it is written, may be subjected to a series of manual checks to help ensure its correctness. Various other sections have discussed the foundations of these checking procedures in detail while other procedures are of the common-sense type that are meant to catch, among other things, clerical or transcribing errors.

Format Check

The format in which the table is to be written is largely a function of the individual decision table preprocessor to be used. The manual accompanying the preprocessor will provide detailed format information. Generally, the format checking should include a check on each of the following:

1. Card column requirements.
2. Header, table end, and option control cards.
3. Designation of the start and end of each rule column.
4. Rule and Stub continuation conventions.
5. Order of cards.

Condition Row Check

Each condition row (considered to include any continuations) should have at least one Y, N, or extended entry. Condition rows which have only don't-care (– or blank) or implicit (* and $) entries cannot participate in the decision making process. If such condition rows are present, they should be investigated.

Action Row Check

Each action row (considered to include any continuations) should contain at least one X entry, an extended entry, or an action sequence number. Any actions with all don't-care (– or blank) entries should be investigated. The check on condition rows and action rows is shown graphically in Figure 14-2.

Frequency Row Check

If a frequency row is used for run time optimization, it should be checked against the preprocessor's conventions for such rows.

Initialization and Other Row Checks

Placement and conventions applicable for the preprocessor to be used should be checked. Also, check whether initialization statements are meant to be executed every time the table is executed or just the first time. Preprocessors vary in their conventions for initialization.

Rule Column Checks

Each rule column should contain the following:

1. A Y, N, or extended entry in at least one of its condition entries.
2. Unless one wants to "fall through" the table, an X, extended entry, or action sequence number in at least one of its action entries. If an action sequence number appears in one of the action entries for a rule, action sequence numbers should appear with every other action of that rule.

This is shown graphically in Figure 14-3.

Stub Column Check

Condition stubs should not start with the word "IF." If extended entries appear in a given condition row the stub should just be a partial statement of the condition such that the extended entries complete the

DEBUGGING PROGRAMS CONTAINING DECISION TABLES

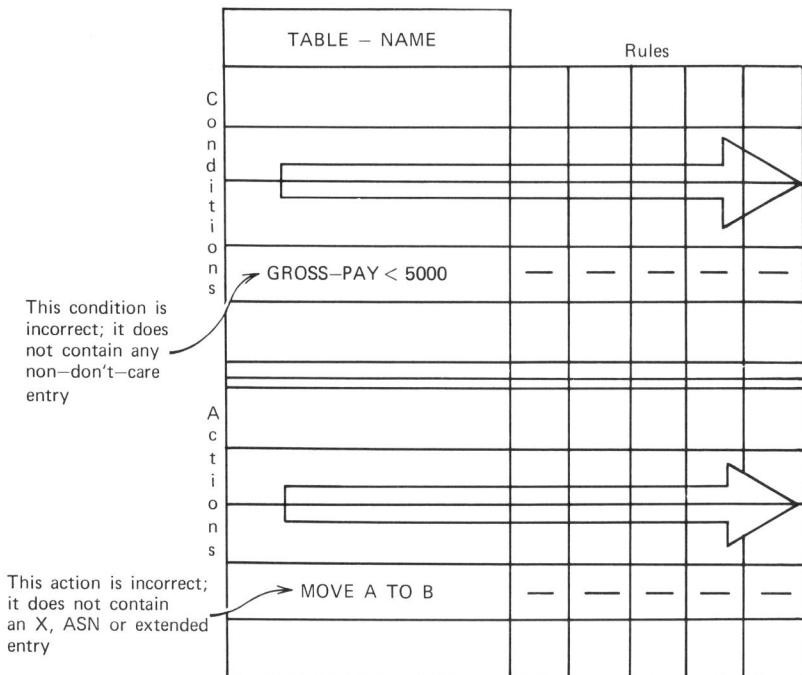

Note: Each condition or action row should contain at least one non-don't-care (and nonimplicit for conditions) rule entry.

Figure 14-2 Check for at least one significant entry in each row (Condition and Action.)

statement. Stub continuation rules applicable to the preprocessor should be checked.

Manual Checking of The Full Table

The preceding manual checking discussion involved parts or components of the table. These parts may be correct in themselves but their logical interactions may be in error. One of the major advantages of decision tables lies in the simple procedure for checking the logical structure represented by the table. Redundant rules as well as any rules which contradict one another are quickly located manually. Of course, all good preprocessors will do the job of redundancy and contradiction checking for you, but the habit of including it in the manual procedure has proven valuable to many users of decision tables.

Checking Procedure

The subject of contradiction and redundancy is discussed in detail earlier. At this point a "cookbook" procedure is presented for manual performance of the check.

Generally, any two rules must be independent[1]; that is, for at least one condition row, one of the two rules must contain a Y, $, or an extended entry while the other rule contains, for the same condition, an N, *, or a different extended entry.

In the event that two rules are not independent, they are either contradictory or

---

[1] It is very important, especially in debugging, to remember the two different ways the word *independent* is used in decision tables. In *rules,* two rules are independent if they are mutually exclusive; while in *conditions* two conditions are independent if they are not related; two mutually exclusive conditions are dependent.

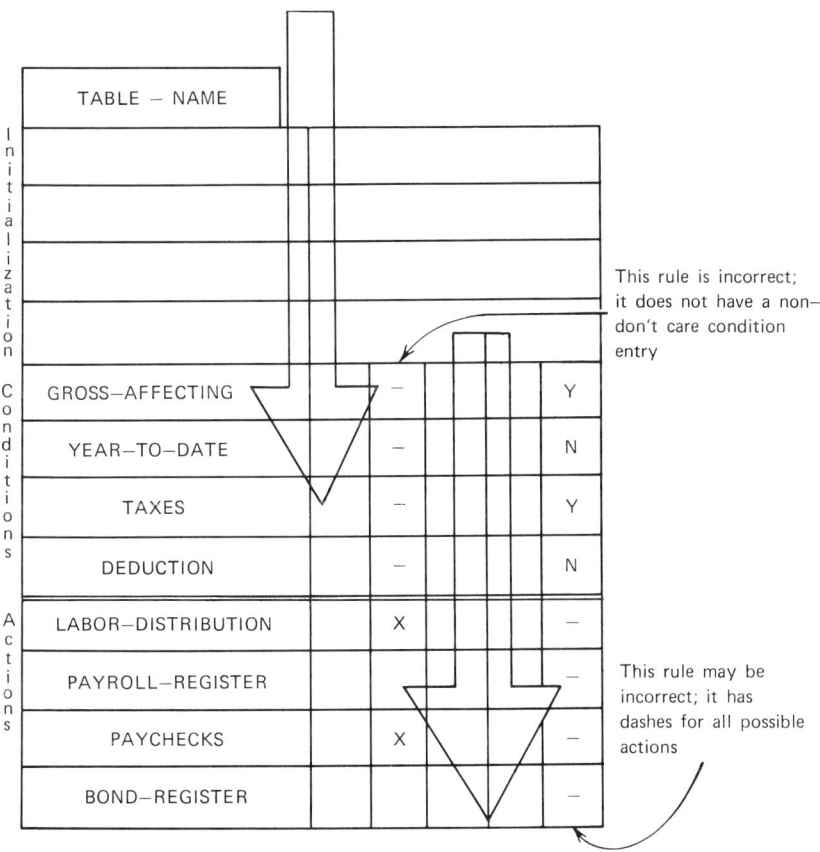

Note: Initialization statements are executed upon every entry to the table (many preprocessors provide exceptions). Each rule should contain at least one non-don't-care (and explicit) entry in the conditions and at least one non-blank entry in the actions.

Figure 14-3 Check for at least one significant entry in condition portion and action portion of each rule.

redundancy exists. If two dependent rules have identical actions, redundancy exists; but if their actions are at all different, they are contradictory.

To do a manual check:

(a) first select the leftmost rule;

(b) compare the selected rule with each of the remaining rules to its right;

(c) look at each condition row in turn, for these two rules only.

   (i) If any one rule's condition entry is Y or $ and the other's is N or * (or if the two rules' condition entries

are different extended entries) then the two rules are independent.

If one or a pair of condition rows represent a range situation (i.e., inequalities are used in the condition stub), independence exists when there is no overlap of the *ranges* being defined, regardless of what the components of these ranges might imply individually. Preprocessors generally do not interpret independence in terms of ranges, so special care should be taken in manual checking of such conditions. Go to Step d.

   (ii) If for the two rules, no condition

row contains differences of the type just described, check the actions of the two rules. If the actions are identical, the rules are redundant; otherwise they are contradictory (see the two following sections, *Procedure for Redundant Rules* and *Procedure for Contradictory Rules*).

(d) Replace the selected rule with the next rule to its right. If this is the rightmost rule in the table (the ELSE-rule is ignored) the check is complete, otherwise go to Step b.

Figure 14-4 graphically follows the above procedure.

PROCEDURE FOR REDUNDANT RULES. So you have found a pair of redundant rules! The first step is to see if an error exists. Redundancy implies that at least one simple rule has been included in both as-written rules. This simple rule could be isolated and thrown out of one of the rules with no harm, but only if the redundancy in fact exists, that is, there has been no error in writing the table. Most preprocessors will handle redundant rules by combining them properly or eliminating the redundancy from one of them. But the user may still wish to do so during the manual checking procedure.

The two sections on *The Rule* discussed simple rules and combination of rules in detail. Here, we give a quick recap of the rule combination procedure.

1. A set of two or more redundant rules may be combined two at a time. Only if the two rules have the same actions and differ in only a single condition row—a Y and an N—will each combination result in a single rule which will itself be redundant with all

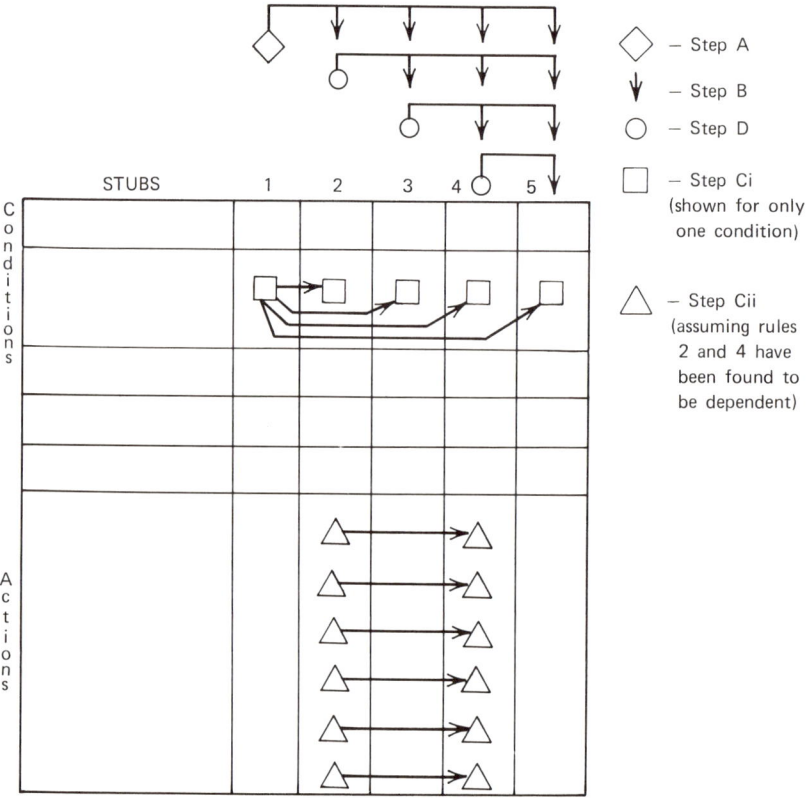

Figure 14-4 A graphical representation of the redundancy/contradition checking procedure given in the text.

THE LANGUAGE, THE HOST PROGRAM, AND THE TABLES

other members of the original set of redundant rules.

2. Two combinable rules may be combined on a condition by condition basis. Only one of three possible cases will be encountered for any given condition.

(a) One rule entry is exactly equal to the other's. The new rule's entry for that condition is the same.

(b) One rule's entry is implicit while the other's is explicit (Y and $, or, N and *). The new rule's entry will be the explicit entry (Y or N). Note that in this case, an error in the assignment of the original implicit entries probably exists.

(c) One or both rule's entry is a don't-care. The new rule's entry is also a don't-care.

3. Because actions are all the same in any set of redundant rules, the actions of the new rule are also the same.

In Figure 14-5 two redundant rules (which cannot be directly combined since they differ in more than one condition) are expanded into their component simple rules. Both expansions contain the simple rule "Y, Y, N."

Removing the duplicated simple rule from first one and then the other rule (but not both) produces these two sets of independent rules of Figure 14-6.

STUBS...	RULES...		STUBS...	RULES...	
C-1	Y	–	C-1	Y	N
C-2	N	Y	C-2	–	Y
C-3	N	N	C-3	N	N
A-1	X	X	A-1	X	X

Figure 14-6

PROCEDURE FOR CONTRADICTORY RULES. Whereas redundant rules *may* represent an error, contradictory rules *definitely* represent an error. The source of this error should be investigated and corrected. After doing so, the manual checking procedure should be repeated; an example of two contradictory rules is shown in Figure 14-7. The rules are further expanded into their simple rules, and the offending simple rule is isolated in that figure.

STUBS...	RULES...			
CONDITION SECTION				
X = 1	Y	–		
Y = 2	–	N		
ACTION SECTION				
GO TO	ED	JOE		
CONDITION SECTION				
X = 1	Y	Y	Y	N
Y = 2	Y	N	N	N
ACTION SECTION				
GO TO	ED	ED	JOE	JOE

Note: The two contradictory rules are expanded into their component simple rules. The case of X = 1, Y Not = 2 is shown to be the one that gives rise to the contradiction and thus is a "bug" to be resolved.

Figure 14-7 Exposure of the contradictory rules in a table.

STUBS...	RULES...	
Conditions		
C 1	Y	–
C 2	–	Y
C 3	N	N
Actions		
A 1	X	X
Conditions				
C 1	Y	Y	Y	N
C 2	N	Y	Y	Y
C 3	N	N	N	N
Actions				
A 1	X	X	X	X
Conditions		
C 1	Y	N
C 2	–	Y
C 3	N	N
Actions		
A 1	X	X

Figure 14-5 Elimination of a Redundant Rule.

Manually Checking the Host Program

Uppermost at this point is ensuring that inter-table communication is correct. Rather than a detailed procedure, a list of questions may be more appropriate.

1. Are all tables to be used (and referred to) in the program present?
2. Is each table treated as though it were a COBOL SECTION? Check here for conventional coding placed after a table such that it might inadvertently become part of the section represented by the table.
3. Have all data items used in the tables been defined?
4. Have all the files referred to in the table been defined?
5. If a naming convention is used, do all data and procedure names—including table-names—agree with the convention?
6. Are the proper control cards present?

Of course host program checking is an entire topic far broader than this short set of questions. Each installation should develop its own check list to be used by all analyst/programmers writing or using decision tables.

Debugging Runs on the Computer

Debugging of computer runs begins after manual checking. Preprocessors have various built-in aids that can best be understood by studying the user's manual for the particular preprocessor. The compiler to be used, the computer operating system, and local conventions will all affect significantly the details of the debugging operation. We now describe a particular procedure to illustrate techniques rather than define procedures.

The Diagnostic Run

Preprocessors are designed to tell you a great deal about the correctness of your tables. All of the points mentioned under manual checking procedures as well as many others may be diagnosed automatically. The tables may be processed separately from the balance of the computer program by the preprocessor just to see what the preprocessor finds wrong with them.

Any errors noted and not corrected automatically by the preprocessor should be corrected in much the same way that errors detected in the manual checking procedures were corrected.

The logical control resulting from decision table usage should greatly ease the burden of debugging during test executions of entire programs. Less emphasis on checking out logical control is needed, so that analyst/programmers can focus their attention on such features as data structure and conceptual correctness.

The latter object, conceptual correctness, points up the fact that the test execution is the last major link in the bridge between concept—how a given program is *intended* to solve a particular problem—and fact— that it *does* solve that problem.

Conceptual correctness can only be verified by finding some other way of solving a sample of the problem and comparing the results obtained. Thus dummy test data may have to be developed. In any event, some way must be found to verify that data is handled as expected.

Compiling the Host Program

Tables may be placed in the host program, preprocessed and then passed to the compiler. Any errors in language (e.g., name definition and syntax) will be diagnosed at this point. They should be corrected with the correcter on the alert for possible introduction of new errors in table logic. This step is repeated until a "clean compile" is obtained.

Where traces are desired (see "Tracing

Flow of Control" later in this section) they should be considered during preprocessing and compilation.

Module Checking

Often programs will call other modules (*subroutines*) from a library. Depending on the operating system, a Link-Edit or similar loading arrangement may produce diagnostics before execution may begin.

The Execution: Test Data

The host program containing the necessary decision tables is now ready to be executed. The tailoring of test input data to cause a systematic testing of the program is common procedure.

Test data read into the program usually determines the flow of control though the program. Various parts of the program are meant to handle particular aspects of combinations of input data while, typically, there is a main control path that is common to all input data.

When decision tables are used, control paths and the data that cause them to be executed are easily identifiable. The first test execution may be done with input data that executes only the main control path. Once this path works correctly, data that exercises other paths leading out from it should be used. As paths, or groups of paths are found to be correct, the data may be designed to exercise finer and finer differences in control path.

EXHAUSTIVE TESTING AND THE "WORST-CASE." Many large programs are so involved and capable of so many possible combinations and interactions of input data that truly exhaustive testing (explicit checking of every possible path of control) is impractical.

One approach to testing such programs is the use of random data. Automatic test data generation programs that permit the creation of such data in volume are available both from manufacturers and software houses. The generation of volumes of randomized data will assure that a wide variety of paths of control and combinations of such paths will be executed during testing. There can be no guarantee that every path or combination of paths will be executed. However, this type of testing has proven valuable.

One more special case of test data is worth mentioning—the *worst-case*. Worst-case analysis has proven very useful in engineering and in programming. It consists of determining a combination of test data that will cause a large number of control paths to be executed and involve maximum or extreme processing in those paths. Such data constitutes the "worst-case." Figure 14-8 shows a simple table and a set of worst-case data items with which to exercise it.

STUBS...	RULES...		
X > 99	*	N	*
X > 3	N	Y	*
X > 1	Y	$	N

*The above decision table (conditions shown only, where the * and $ entries express dependencies within the conditions) can be exercised by a typical worst-case set of data below (assume integers only):*

X = 100	(barely within the Else rule)
X = 99	(high extreme of Rule 2)
X = 3	(high extreme of Rule 1)
X = 1	(Rule 3)

Figure 14-8 Sample table and worst-case data to test the table.

Error Detection and Recovery Checks

Errors in input must be detected by the program which must, in turn, *recover* its performance as nearly as possible after the error detection. How extensive error-detection and error-recovery procedures are to be is determined earlier in the programming or analysis stages, but tests to determine if such detection or recovery procedures operate as planned are necessary parts of the debugging stage. Error cases may be individually designed, included in the random test data or deferred until debugging is nearly complete.

Tracing Flow of Control

One of the more useful debugging tools is the decision-table trace. This feature creates a record indicating the name and rule number of the selected rule every time a table is entered. The record enables the logical flow of the program to be examined.

A typical application will pass a series of records (*transactions*) against a table or set of tables. If the record contents are exhibited as each is read, the logical path through the program associated with that record may be recorded with the help of the decision table trace.

Coding outside of decision tables may be traced by conventional COBOL means. In some preprocessors, the automatic trace is synchronized with DISPLAY, EXHIBIT, and TRACE feature of IBM's S/370 COBOL compilers so that these features can be combined to produce a true flow of control record.

Where a particular decision table preprocessor does not offer a trace feature, it can easily be simulated by including a DISPLAY of the table name and rule number as the first action of each rule.

Production Testing

Testing must continue even after the program is thoroughly "wrung out" and deemed ready to go into production.

If the new program is meant to replace an older program or a manual procedure, the *parallel run* approach may be used. For a period of time all data to be processed by the new system is also processed by the old system. This redundant approach allows a close check on the results obtained and comparisons of costs involved. Most importantly it protects the using organization against the expensive situation of an undetected error.

As the level of confidence in the new system rises, the old system should be discontinued. As may sometimes happen, however, the level of confidence in the new system may nosedive and it will be the new system rather than the old which will require scrapping.

As the new program becomes accepted, it leaves the debugging stage. But it very likely is not 100% error-free. From time to time previously undetected errors will crop up, or the problem that the program is solving will change. This new stage that the program has entered is known as *maintenance*. The joys of maintenance are discussed in the next section.

SECTION 15

Maintenance of Decision Tables

Once a program is in production one of two events may occur: (a) previously undetected errors may crop up and require correction or (b) the problem that the program was meant to handle may change.

Handling of these two situations is the province of program maintenance. Generally, the maintenance programmer is not the same individual as the programmer who originally wrote the program, and the maintenance activities are required anywhere from months to years after the original programmer's job is done.

Because of these gaps in time and people, communications and documentation become major factors in successful maintenance procedures. The use of decision tables in programming is a significant ingredient in aiding communications between the original implementor of the program and the later maintenance programmer—two people who may never meet.

Because the original implementor's role in this communication process is especially important, maintenance must not begin with the transition from debugging to production—it must be considered from the earliest stage of program conception.

To illustrate maintenance procedures for handling changes to programs that contain decision tables, we use the simple Bond Deduction Register program described in Section 12 (*Decision Tables in Computer Programs*). That table is duplicated in Figure 15-1 in a shortened fashion.

	STUBS...	RULE ENTRIES...			
C	AT-END-SW = 1	N	N	N	Y
C	DEPT NOT = DEPT-OLD	N	N	Y	–
C	DED-BOND = 0	N	Y	–	–
A	PERFORM PLACE-ON-REGISTER	X	–	–	–

Figure 15-1 Simplified decision table for bond deduction register.

SPLITTING RULES

Assume that we need to make a slight change to the Bond Deduction Register program so that it now excludes a specific department, say Department 13, from the bond register. Perhaps that department has a special bond purchase arrangement differing from that of the other departments; the reason for this change is immaterial for purposes of this example.

The rules previously represented in the table were the following:

Rule 1. Records with bond deductions (DED-BOND NOT = 0).
Rule 2. Records with no bond deductions (DED-BOND = 0).
Rule 3. New department break.
Rule 4. End of record file.

The maintenance programmer now wishes to alter the table to represent five rules:

Rule 1. Records with bond deductions but not for Department 13.
Rule 2. Records with bond deductions for Department 13.
Rule 3. Records with no bond deduction.
Rule 4. New department break.
Rule 5. End of record file.

Notice that the old Rule 1 of Figure 15-1 is split into two new rules, 1 and 2, in Figure 15-2. The difference between these rules is that Rule 1 now has a Y entry for a newly added Department 13 condition, while Rule 2 has an N entry for that new condition. All other rules have a don't-care entry for that new condition since it doesn't enter into their considerations.

This example of decision table maintenance illustrates splitting a rule for addition of a new condition. Whenever a rule is split into one or more new rules, those new rules are automatically independent from all of the other old rules (assuming the original rule was independent before splitting, of course) and the maintenance programmer need only concern himself with independence within the new rules created from the old one.

In the example of Figure 15-2, the newly added condition, with an N for the new Rule 1 and a Y for the new Rule 2, serves to ensure independence.

Rule combination is treated much as it is in writing a table—any two rules having the same actions and having in one condition row a Y, N (or equivalent) pair for limited entry, or different values for extended entry, may be combined. A more complete

	STUBS...	RULE ENTRIES...				
C	AT-END-SW = 1	N	N	N	N	Y
C	DEPT NOT = DEPT-OLD	N	N	N	Y	–
C	DED-BOND = 0	N	N	Y	–	–
C	DEPT = 13	N	Y	–	–	–
A	PERFORM PLACE-ON-REGISTER	X	–	–	–	–

Figure 15-2 Decision table of Figure 15-1 with additional condition.

	STUBS...	RULE ENTRIES...				
C	AT-END-SW = 1	N	N	N	N	Y
C	DEPT NOT = DEPT-OLD	N	N	N	Y	-
C	DED-BOND = 0	N	N	Y	-	-
C	DEPT = 13	N	Y	-	-	-
A	PERFORM PLACE-ON-REGISTER	X	X	-	-	-

⇩

C	AT-END-SW = 1	N	N	N	Y
C	DEPT NOT = DEPT-OLD	N	N	Y	-
C	DED-BOND = 0	N	Y	-	-
A	PERFORM PLACE-ON-REGISTER	X	-	-	-

Figure 15-3   Return of Figure 15-2 to Figure 15-1.

discussion of rule combination is included in the section dedicated to this subject.

Rules 2 and 3 of Figure 15-2 have the same actions but differ in more than one condition, thus they are not eligible for combination. But let us now assume that the maintenance programmer, a year after changing the table of Figure 15-1 to that of Figure 15-2, is informed that Department 13 is now once again to be included in the bond deduction register. He must now change the actions of Rule 2 to be identical to those of Rule 1 in order to be listed in the register. Figure 15-3 shows the combination of the rules.

Rules 1 and 2 (in the top portion of Figure 15-3) now have the same action and differ in only one condition. These two rules may be combined, the Y and N combining into a don't-care entry. Now the Department 13 condition row contains only don't-care entries, and may be removed from the table. The result (in the lower portion of Figure 15-3) becomes identical to Figure 15-1.

Communications Aspects of Rule Combination

If two independent rules are eligible to be combined, is it always necessary to actually do the combination? Technically, no. No preprocessor will require you to combine rules; some will even combine the rules for you. The main criterion for deciding whether or not to combine rules is communications—someone, at some later date may have to look at this table and want to understand and change it. The rules should have some logical meaning and some communications capability of their own.

To combine two rules that really don't have much to do with one another, just because they happen to have the same actions and are combinable, may hurt the cause of communication. Likewise, to leave several obviously related and combinable rules separate on general principles may make the table seem more complex than it really is.

Adding New Rules

We are speaking here of adding new rules to be based on the set of conditions already stated; some comments on adding new conditions appear later in this section.

If the rule being added is already included in an existing non Else-rule, the problem reduces to splitting that rule, as previously discussed. If an entirely new rule is being added, it amounts to extracting it from the ELSE-rule.

The ELSE-rule, as discussed in the section on *Completeness Checking,* represents all of the simple rules not represented by the other written rules in the table. It is from this reservoir of the ELSE-rule that completely new rules (i.e., not redundant or contradictory to any existing rule) must be drawn. All of the existing rules will remain independent from one another but the new rules must be checked for the following situations:

1. The set of all rules must be independent (i.e., no contradictions or redundancies) from one another.
2. Each new rule must be independent from each old rule.

CHANGING A CONDITION

A condition may be split into several conditions, or several conditions may be combined into one. Furthermore, a condition may be changed from limited to extended entry or vice versa. The principal considerations of these changes are those having to do with rule independence and condition dependence. The earlier sections on conditions and rules contain extensive discussions of these effects that will not be duplicated here. However, we point out some of the more common problems encountered in maintenance.

We remind the reader of the differences in the use of the word "independent" in connection with conditions and rules. Two conditions are independent if there is no predictive effect of one on the other—knowing one condition's truth status tells you nothing of the other's status.

Rules, on the other hand, are independent when they are mutually exclusive (mutually exclusive conditions are dependent, not independent). Thus if one rule is selected, it automatically follows that no other rule can be selected at the same time. Rules that are not independent (mutually exclusive) are either redundant (have the identical series of actions) or contradictory (have different actions).

Changing a condition may have effects on either, or both, condition independence and rule independence. Since a given condition may be responsible for the independence of two or more rules, it is best to do a complete redundancy/contradiction check (see the section on *Debugging*) on the entire table after making the change.

Adding a New Condition

Adding a new condition to a table will not cause loss of independence of rules. However, it will serve to further limit the number of simple rules represented by the as-written rules (see *Completeness Checking*). The implication here is that additional simple rules may have been diverted to the ELSE-rule. If no ELSE-rule was required before the addition of the new condition, one may well be required afterwards.

Often, a new condition is added in conjunction with a new rule; Figure 15-2 showed such a case. The considerations of the new rule, being more extensive than those for a condition row, take priority in such cases.

Removing a Condition

Removing a condition may have an effect on both rule independence and completeness. Whenever a condition is removed, a complete manual checking procedure (see *Debugging*) should be performed. However, a condition row containing only don't-care entries may always be removed with no adverse effects on the table.

ACTIONS

Actions associated with a set of rules in a table may be changed with a great deal of freedom. However, the following checklist should be referenced when making changes:

1. Should extended or limited entry format be used? Communication value of the table as a document rather than as part of a program is the determining consideration.
2. If the original table had any redun-

dant rules (i.e., no Y, N type difference between them in the conditions) care must be taken to keep them from becoming contradictory.

3. Rules which contain don't-care in their conditions represent more than one simple rule (see *Completeness Checking*). Before making an action change to such a rule, verify that the change really applies to all of the simple rules involved. If the action change is only applicable to some of these simple rules but not to others, the rule must be split, as discussed earlier.

4. Sequencing of actions may be changed. Action sequence numbers, if permitted by the particular preprocessor to be used, may be used to express, for each rule individually, the order in which the actions are to be executed. Where action sequence numbers are not allowed, separate sets of action rows, each set in appropriate sequence, can be associated with the proper rules.

5. Rule combination procedures may be investigated on completion of the action changes, in order to simplify or improve the documentation value of the tables.

MISCELLANEOUS MAINTENANCE CHANGES

Other aspects of tables or of the organization of tables within the host program may require maintenance changes. Without speculating on the many situations that might require such changes, we now discuss the effect of various changes upon a table or program.

Initialization Changes

Initialization statements are executed every time a table is entered, whether it is being performed, referenced by a GO TO or entered as a loop from within itself. If initialization is desired only once rather than every time the table is entered, such initialization can be accomplished in several ways. It can be performed outside of the table, or with the help of a first-time switch (which must itself often be initialized), or the initialization statements may be made into actions associated with a separate rule that controls when the initialization is to be done.

Increasingly, preprocessors are offering various ways of entering tables so that initialization may be bypassed. These include a LOOP verb, used as an action-row statement, to re-enter the table in which the action is located; a GO TO some special form of the table-name; definition of a special dummy rule to contain initialization (executed only on the first entry to the table), and so on. Each of these has disadvantages and DECITAB provides no special means of entering the table in order to bypass initialization.

Run Time Optimization and Frequencies

During the life of a program, the distribution of the data may change so that improved performance may be realized by introducing a run time optimization technique, or changing frequencies if such optimization is already being utilized. Generally, the changing of relative frequencies of the rules of a table may be done freely. However, such changes in relative frequencies may cause the size of the program to change. Removal of frequencies may be desirable if minimizing the amount of required storage becomes more important than minimizing run time. The section on optimizations contains more information on this subject.

Adding, Deleting, and Restructuring Tables

These operations are major surgery at times, comparable to (or worse than) developing original tables and programs. Fitting one major logical structure into another, or removing such a structure, requires knowledge not only of what the final logical structure will look like, but also of the existing structure's details. The section on *Decision*

Table Interactions discusses this point more fully.

Grouping of Actions

The sets of actions executed by two different rules may have common members. When making maintenance changes the relationships of these common members may change. Optimization of common actions between two or more rules proceeds in maintenance much as it does in any other case. The section on *Optimization* discusses this point in detail.

Changing Table Names

Table names generally are tied to the rules of the language used; COBOL-oriented preprocessors will treat table names as section names as defined by COBOL and by the compiler to be used. When tables are restructured, all references to the original table name must be appropriately changed, with the degree of uniqueness required by the particular compiler considered.

DOCUMENTING MAINTENANCE CHANGES

Documentation of maintenance changes varies from scribbled notes on backs of envelopes to elaborate reporting schemes such as the IBM Corporation's APAR (*Applied Programming Analysis Report;* this is one of the more common names fitted to those letters, whose original meaning has been lost with time).

A summary of recognized techniques that seem to perform well for documenting maintenance changes is as follows:

1. Every problem or change to a table or program is identified with a unique number.

2. A priority system ranks the relative importance of various changes so that the programmer's time can be channeled to where it will do the most good.

3. As changes are made, a record of each change is made and identified by the number assigned above.

4. The new or modified table is run through the preprocessor and the new listing of the table and its decomposition replaces the listing of the original table and its decomposition.

By documenting specific changes to a program in a formalized manner, control over the operation and status of a program is greatly simplified. A sample form, usable for recording maintenance problems, is shown in Figure 15-4.

| Severity
Code | XYZ Corporation
Problem Analysis Report | PAR
No. |

System in which problem noted _____

Subsystem or Program _____ Date _____

Table Name _____

How Serious? _____ Reporter _____

(Check one)

    Production System cannot proceed   _____

    Other System cannot proceed   _____

    Minor problem   _____

    Potential, not currently, troublesome   _____

    Improvement or Suggestion   _____

Brief Description (use additional pages for detailed description)

_____

_____

_____

Please attach listings, notes or any other supporting evidence.

Below for maintenance group use only

Technical Description of Problem _____

_____

_____

Comments   (Initials, date)

                                                    OK – Develop fix   _____

                                                    OK – Test fix   _____

                                                   OK – Fix Tested   _____

                                                   OK – Integrated Test   _____

Solution Dates: _____.   OK – Fix Issued   _____

Maintenance level _____ Version _____.   OK – PAR Closed   _____

Figure 15-4   A sample maintenance report form.

SECTION 16

Decision Table Interactions

LOGICAL STRUCTURES OF DECISION TABLES

Two decision tables are sometimes better than one. In fact, in some logical problems, entire trees of decision tables may be required. The building of logical structures with decision tables as their basic elements can be justified on the basis of the following parameters:

1. Time dependence.
2. Documentation value.
3. Capacity of a single decision table.

Time Dependence

The conditions of a decision table are evaluated at a particular point in time—when the table is executed. Except for the initialization statements, a table can not modify any of its condition-making criteria in a particular execution based on any of the rules that might be selected during that execution.

To illustrate time dependence, consider this simple problem:

> *If the sum of an employee's previous year-to-date FICA deduction plus this-pay-period's FICA deduction exceeds n dollars, calculate this-pay-period's FICA by procedure A, else calculate it by procedure B.*

This problem asks us to calculate FICA, based on the value of FICA. Obviously, we cannot determine the procedure for calculating FICA until after we have already calculated it! But, if we know that procedure B always calculated a FICA deduction equal to or greater than that calculated by procedure A, we can calculate by procedure B first, test the sum and, if the sum is greater than the limit n, recalculate by pro-

116 DECISION TABLE INTERACTIONS

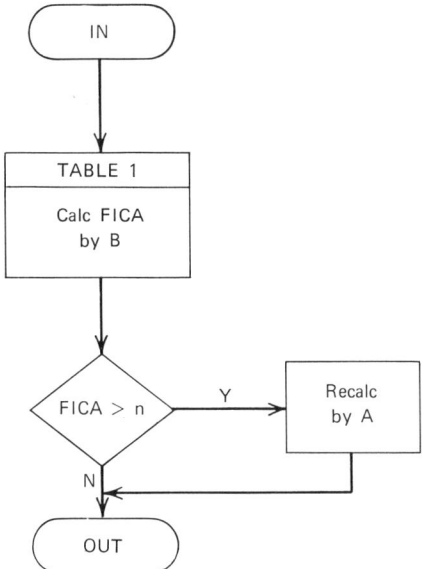

Figure 16-1 Flowchart of sample FICA processing.

cedure A. A flowchart of a structure of tables to do this (the exact contents of each table are not shown here) is presented in Figure 16-1.

Inasmuch as the decision tables we deal with have no provision for describing rule priority, Procedure B cannot appear in the decision table that calls for Procedure A. Furthermore, the processing of the table that uses Procedure B will have to precede the processing of the table that uses Procedure A.

This FICA problem could perhaps more easily be handled by any one of a number of alternate approaches, but it does illustrate a common type of time dependence.

Documentation Value

One of the major advantages of tables is their value as documentation. The solution to a complicated logical problem is presented more clearly and in an organized manner by decision tables. Structures of the solution and their use will greatly facilitate any later maintenance efforts devoted to the program. The many advantages of decision tables over flow charts and narrative for describing complex logic have been described in the literature of decision tables.

Capacity

Every decision table has limits placed on its maximum size. It could well be that the number of conditions, or actions, or rules required in using tables to solve a particular problem might exceed these limits, so as to require the problem to be carefully structured by a system analyst. Each part of the structured system becomes a separate table or substructure of tables.

The exact limitations of the capacity of a decision table are a function of the particular preprocessor to be used in decomposing the decision tables. Refer to your preprocessor's user's manual or documentation for the limiting figures applicable to it.

SECTION CONSIDERATION

Decision tables that are to be converted to COBOL represent COBOL SECTIONs with the table name being the SECTION name. They follow all of the rules of SECTIONs including those related to PERFORM. Paragraph-names within SECTIONs that are developed from decision tables are generally inaccessible to the user. The initialization statements within the table are executed every time the section is referenced even when referenced from within the table itself.

Looping in a Table

Tables may be iterative—that is, a table may contain an action that goes to its own table name. Such a table changes (via actions) some variable which is reevaluated (via conditions) to select a rule. This situation, known as *looping,* is useful in such

MONITORING AND CONTROL TABLES

applications as approximating a function by modifying it based on its most recent value, in searching and summing.

At least one rule must permit either "fall through" or exit from the table via a GO TO a name outside of the table. A looping table is used as an illustration in the section *Systems Analysis*. Initialization statements in looping decision tables are executed every time the table is entered. If initialization is desired before looping rather than during it, the table writer must arrange for it to be done prior to the first entry to the table.

Several decision tables preprocessors have implemented a special action known as LOOP which permits a table to be reentered bypassing the initialization statements.

MONITORING AND CONTROL TABLES

Going beyond the solution of the original problem, control of the solution process becomes a problem in itself. Programs (or decision tables) to control other programs (or decision tables) are called *monitors*.

A monitor decision table might be likened to a macro flow chart—it presents the

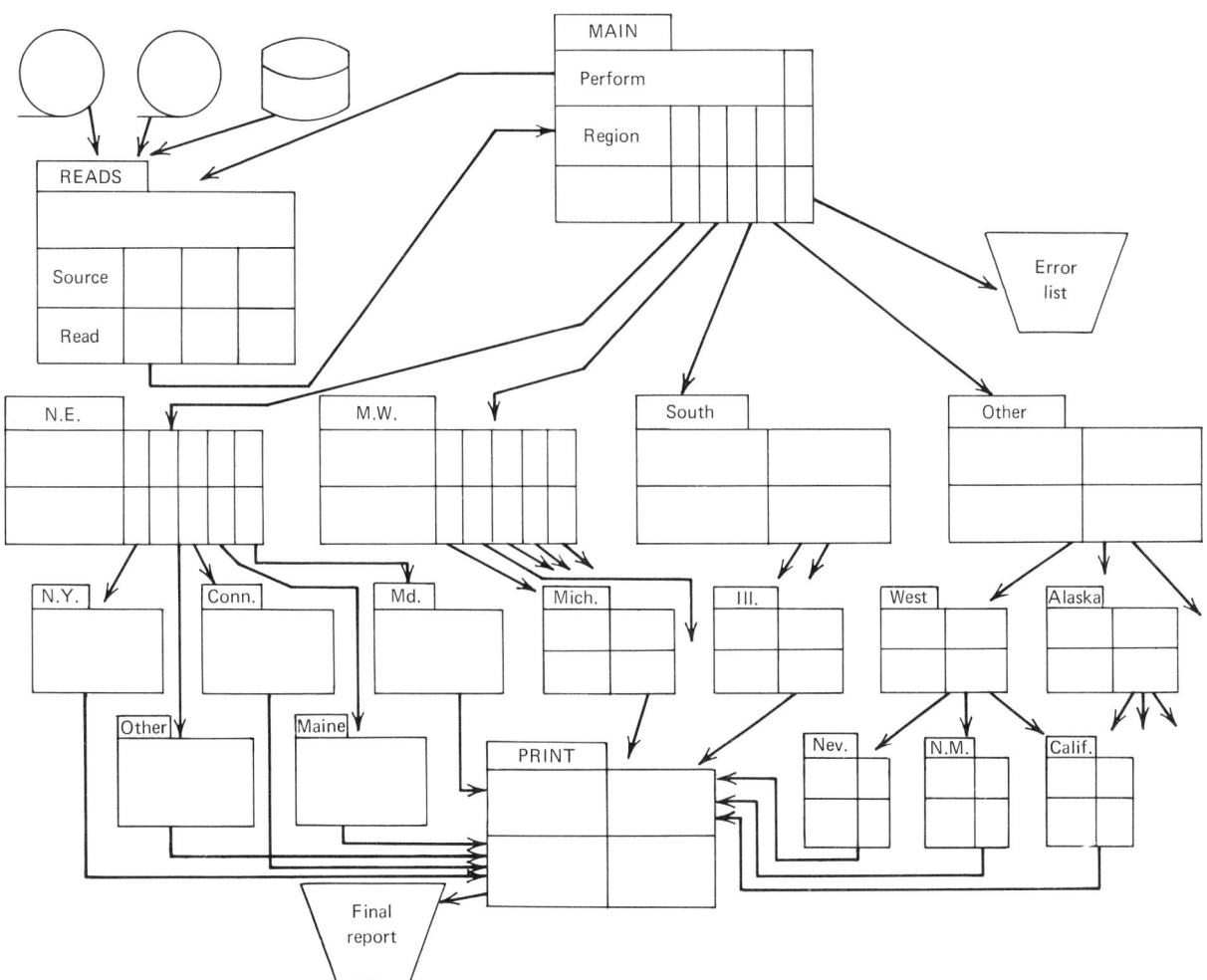

Figure 16-2 **A structure of tables.**

logic of problem solution in very broad terms. This logic is expressed as a set of major conditions that determine the overall operation of the system and by a set of actions that cause other decision tables to do the detail work involved in each of the broader steps.

To illustrate, consider a simple sequential file search for records with various states of origin of products in order to establish routings to a central receiving point.

As illustrated in Figure 16-2, the main (or monitor) decision table senses records to be read via one subordinate decision table, then performs the proper subordinate table depending on the region. Each region processing table performs the proper state processing table. Each of these, finally, performs a single report preparation table.

SECTION 17

A Case Study: Development and Decomposition of a Decision Table

This is a case study of a fictitious systems project that includes a relatively logical programming problem using decision tables. The project was the following:

> *Automate the distribution of customer records to two clerks. The distribution clerks' supervisor will specify the distribution rules.*

This fictitious project is much simpler than those encountered in work-a-day system analysis, but it will serve to illustrate many decision table features.

The distribution supervisor's written description to the system analyst is shown in Figure 17-1. The system analyst looked over this sheet and extracted the conditions and actions resulting therefrom as shown in Figure 17-2. By just looking at Figure 17-2, it becomes obvious that there are serious lapses in the logic shown. There are questions that need to be asked, but which ones? At first the analyst thought that the supervisor's note about "aren't divided up more evenly" referred to the fact that Clerk 1 only got 8 last initials (A–H), while Clerk 2 got 18 (I–Z).

But then, the advertising manager reminded him that their sales came mainly from a single direct-mail ad campaign where the names came from the telephone book—and only covered A through M of the book, so the vast majority of the customers' names started with A through M!

The analyst now had to present in a clear and concise form to the supervisor those cases that were not clearly decided. A

120 A CASE STUDY: DEVELOPMENT AND DECOMPOSITION OF A DECISION TABLE

> FROM: J. Knox (Supervisor, Distribution Department)
> TO: R. Mendel (System Analyst)
> SUBJECT: RECORD DISTRIBUTION PROCEDURE
>
> In accordance with your supervisor's request, I offer the following description of our current record distribution procedure:
>
>> If customer's last name starts with A through H, we send record to Clerk 1, but if customer's last name starts with I through Z we send it to Clerk 2. The reason these aren't divided up more evenly is that Clerk 1 also handles customers in Nevada, and Clerk 2 also handles all customers with over 20 transactions in the previous year. We have found that the workload evens out overall.
>
>> Sorting clerks currently make the distribution of records to the two clerks. Assuming we had a device that could read key information like customer's last name, state location, and customer's total number of transactions for last year, I would like your ideas on automating the distribution of records to the two clerks.

Figure 17-1 Information supplied to system analyst by user.

thorough logical analysis of the situation shown in Figure 17-2 was needed.

After rewriting Figure 17-2 as a decision table (Figure 17-3), the analyst made a complete redundancy and contradiction check. The results are indicated in Figure 17-3. This table could be taken to the supervisor, with the analyst pointing out the contradictions and redundancies and asking the supervisor to resolve them. However, it is preferable for the analyst to pose specific questions for him. To develop these questions, the analyst remembered that a contradiction or redundancy between two rules represents one or more simple rules shared between these rules. So, expanding all of the rules involved in either contradictions or redundancies into their simple rule form, (in this case, all four rules required expanding), he carefully checked for any simple rules which occurred two or more times (See Figure 17-4). While he was checking for simple rules, he also checked the expanded table of Figure 17-4 to make sure that every combination of Y and N for the three conditions was present; they were,

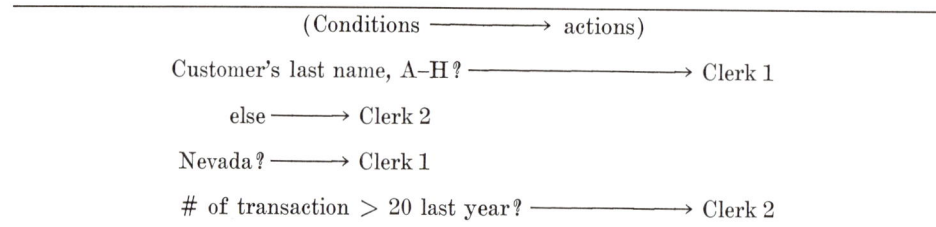

Figure 17-2 Translation of narrative to decision table (Step 1: extracting conditions and actions from the narrative).

ERRATA   JULY, 1971

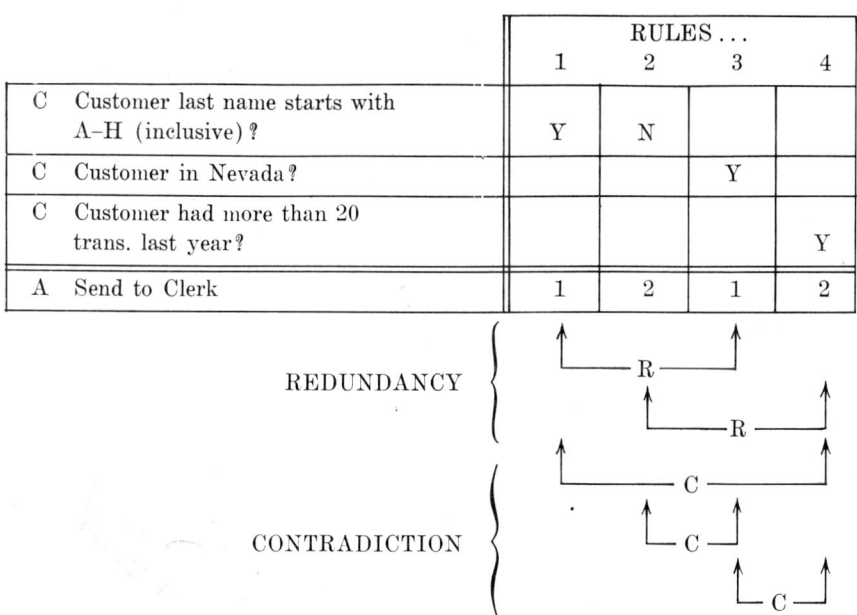

Figure 17-3 Translation of narrative to decision table (Step 2: develop preliminary table and determine contradictions and redundancies).

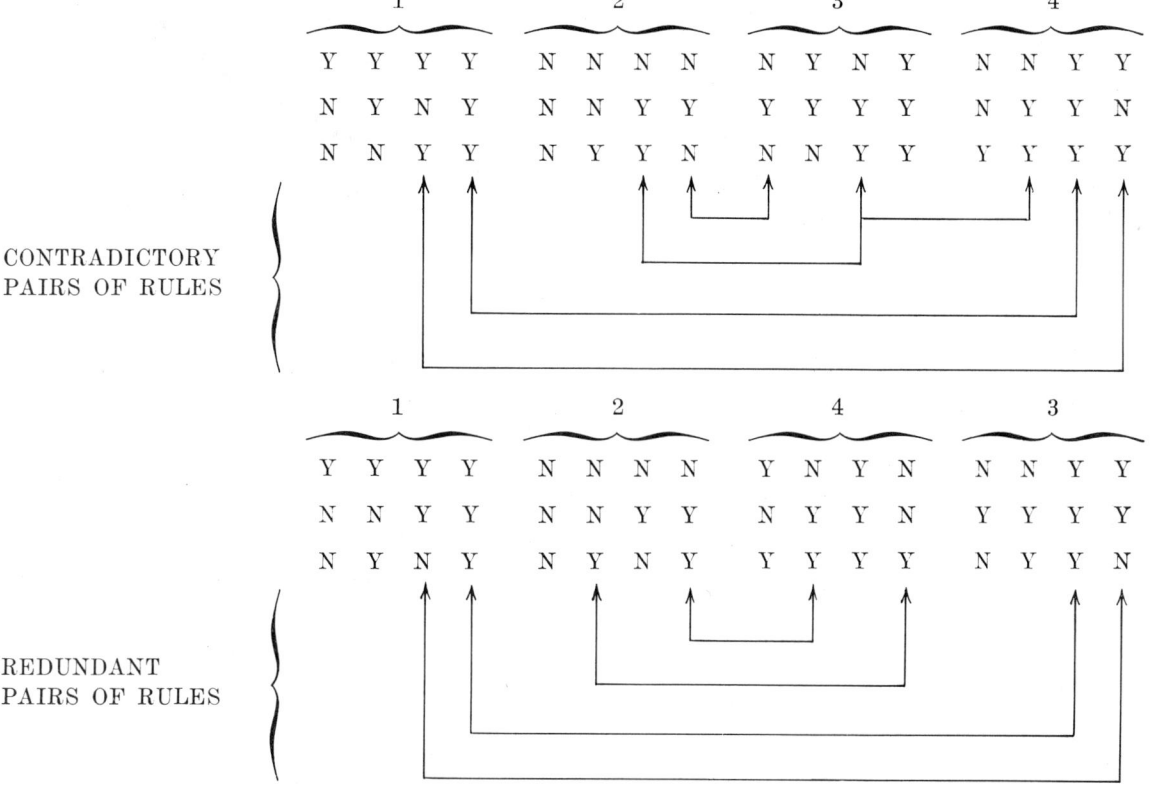

Figure 17-4 Translation of narrative to decision table (Step 3: expanding contradictory and redundant rules).

Pollack et al.: Decision Tables

A CASE STUDY: DEVELOPMENT AND DECOMPOSITION OF A DECISION TABLE

		RULES...			
		1	2	3	4
C	Customer last name starts with A–H (inclusive)?	Y	N		
C	Customer in Nevada?			Y	
C	Customer had more than 20 trans. last year?				Y
A	Send to Clerk	1	2	1	2

REDUNDANCY
- R (rule 3)
- R (rule 4)

CONTRADICTION
- C (rule 3)
- C (rule 4)
- C (rule 4)

Figure 17-3 Translation of narrative to decision table (Step 2: develop preliminary table and determine contradictions and redundancies).

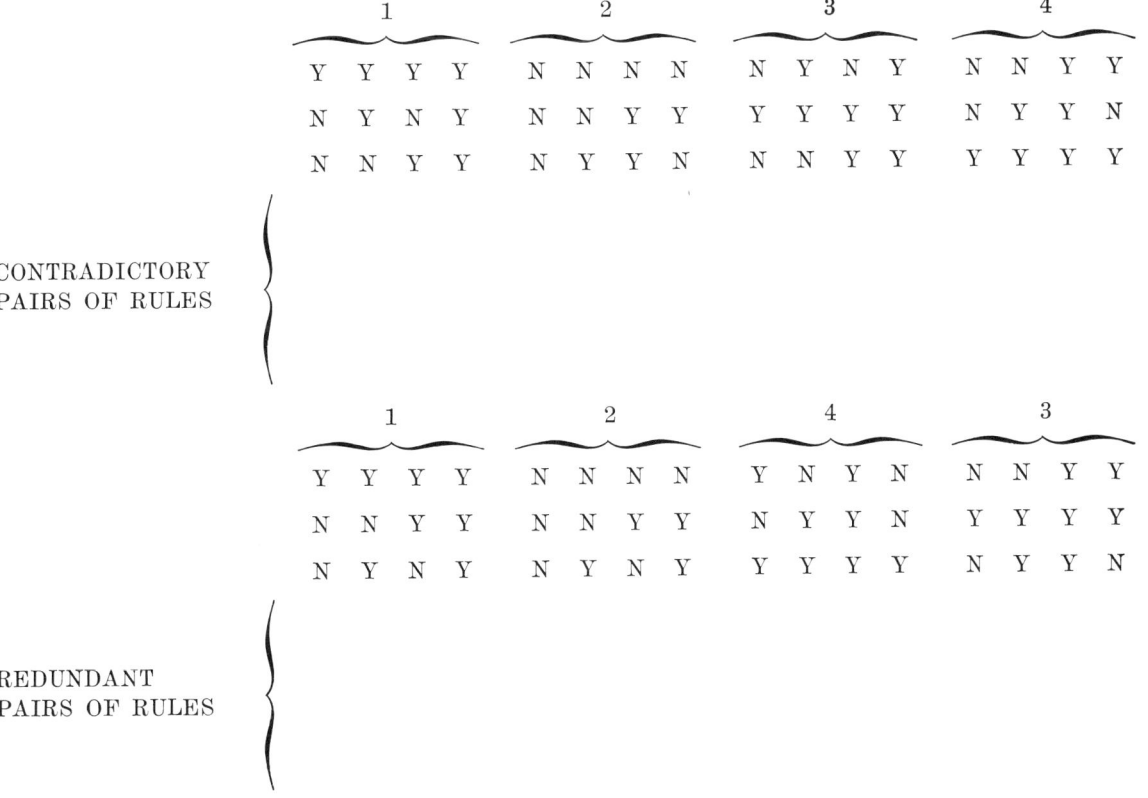

Figure 17-4 Translation of narrative to decision table (Step 3: expanding contradictory and redundant rules).

	CONTRADICTIONS				REDUNDANCY	
Customer last name starts with A–H (inclusive)?	Y	N	N	Y	Y	N
Customer in Nevada?	N	Y	Y	Y	Y	N
Customer had more than 20 trans. last year?	Y	N	Y	Y	N	Y

Figure 17-5 Translation of narrative to decision table (Step 4: delineation of the contradictory and redundant simple rules).

thus his table was complete. At least the supervisor hadn't missed any combination for the three conditions.

All the simple rules in Figure 17-4 that have one or more lines pointing at them (after eliminating duplicate rules) constitute the set of specific questions to be asked of the supervisor. These are shown in Figure 17-5. The following questions were framed along with the supervisor's answers:

1. Initial A–H, not Nevada, and over 20 transactions last year?   1.
2. Initial I–Z, Nevada, and not over 20 transactions last year?   2.
3. Initial I–Z, Nevada, and over 20 transactions last year?   1.
4. Initial A–H, Nevada, and over 20 transactions last year?   2.
5. Is Clerk 1 right for A–H, Nevada, and not over 20 transactions?   yes.
6. Is Clerk 2 right for I–Z, not Nevada, and over 20 transactions?   no.

The supervisor pointed out in this case, that for a good customer (Krolls) in Arizona, Clerk 1 would take it; otherwise, yes, Clerk 2 gets it. Using this new information the system analyst made up the new decision table shown as the top table in Figure 17-6. The asterisks have been introduced to represent existing condition dependencies.

The lower table in Figure 17-6 is the system analyst's final product, created by combining the rules of the upper table.

With this table, the programmer need only determine the proper programming tests to be made in the condition stubs and programming actions to be taken in the actions section, and the program in which to fit the table. A very simple host program is shown in Figure 17-7, with its COBOL decomposition (procedure only), developed by the algorithm given in the chapter on decomposition, shown in Figure 17-8. The bounded-operator is used in the programming table for efficiency.

A CASE STUDY: DEVELOPMENT AND DECOMPOSITION OF A DECISION TABLE

C	Customer last name starts with A–H (inclusive)?	Y	Y	N	N	N	Y	Y	N	N
C	Customer in Nevada?	N	N	N	Y	Y	Y	Y	N	N
C	Krolls in Arizona?	–	–	–	*	*	*	*	N	Y
C	Customer had more than 20 trans. last year?	N	Y	N	N	Y	Y	N	Y	Y
A	Send to clerk #	1	1	2	2	1	2	1	2	1

(The 1,1 pair collapses to R1; the 2,2 pair collapses to R2)

		R1	R2	R3	R4	R5	R6	R7
C	Customer last name starts with A–H (inclusive)?	Y	N	N	Y	Y	N	N
C	Customer in Nevada?	N	–	Y	Y	Y	N	*
C	Krolls in Arizona?	–	–	*	*	*	N	Y
C	Customer had more than 20 trans. last year?	–	N	Y	Y	N	Y	Y
A	Send to clerk #	1	2	1	2	1	2	1

LOGICAL INTEGRITY CHECKING

	R1	R2	R3	R4	R5	R6	R7	
Simple Rule Counts:	2^2	2^2	2^1	2^1	2^1	2^0	2^0	
Sum	= 4	+ 4	+ 2	+ 2	+ 2	+ 1	+ 1	= 16

Note: For rule-counting purposes, the *'s of Rules 3-5 are considered as dashes. The * of Rule 7 is not considered as a dash inasmuch as its counterpart has already been included in Rule 3.

Number of simple rules possible = 2^n, n = number of conditions = 4

$$2^4 = 16$$

There are no redundancies or contradictions in this table (i.e., any two rules will have at least one Y, N pair on the same condition). Thus the table is complete and is logically sound.

Figure 17-6 Translation of narrative to decision table (Step 5: development of corrected table plus logic integrity checking of that table).

A CASE STUDY: DEVELOPMENT AND DECOMPOSITION OF A DECISION TABLE

```
OPEN-UP SECTION.
OPEN-UP-1.
    OPEN INPUT CUSTOMER-FILE, OUTPUT CLERK-FILE.
READ-IT.
    READ CUSTOMER-FILE AT END GO TO CLOSE-UP.
DECITAB  ASSIGN-CLERK
```

		RL-NR	1	2	3	4	5	6	7
C	LETTER-1 < I		Y	N	N	Y	Y	N	N
K	LETTER-1 IS INITIAL								
C	STATE = 'NEV'		N	–	Y	Y	Y	N	*
C	CUST-NAME = 'KROLLS OF ARIZONA'		–	–	*	*	*	N	Y
C	TRANS (LAST-YR) > 20		–	N	Y	Y	N	Y	Y
K	ACTIVITY								
A	COMPUTE CLERK =		1	2	1	2	1	2	1
A	WRITE CLERK-RECORD		X	X	X	X	X	X	X
A	GO TO READ-IT.		X	X	X	X	X	X	X

```
TABEND
CLOSE-UP SECTION.
CLOSE-UP-1.
    CLOSE CUSTOMER-FILE, CLERK-FILE.
    STOP RUN.
```

Note: K in Column 12 signifies a comment.

Figure 17-7 Host program (including decision table).

A CASE STUDY: DEVELOPMENT AND DECOMPOSITION OF A DECISION TABLE

```
OPEN-UP SECTION.
OPEN-UP-1.
    OPEN INPUT CUSTOMER-FILE, OUTPUT CLERK-FILE.
READ-IT.
    READ CUSTOMER-FILE AT END GO TO CLOSE-UP.
ASSIGN-CLERK SECTION.
Q1.
    IF LETTER-1 < 'I' GO TO Q2.
    IF TRANS (LAST-YR) > 20 NEXT SENTENCE
        ELSE GO TO RULE-02.
    IF STATE = 'NEV' GO TO RULE-01.
        (Note: Rule-03 has same actions as Rule-01).
    IF CUST-NAME = 'KROLLS OF ARIZONA' GO TO RULE-01
        ELSE GO TO RULE-02.
        (Note: Rule-07 has same actions as Rule-01; Rule-06 same as Rule-02).

Q2. IF STATE = 'NEV' NEXT SENTENCE ELSE GO TO RULE-01.
    IF TRANS (LAST-YR) > 20 GO TO RULE-02.
        (Note: Rule-04 has same actions as Rule-02, Rule-05
            is drop-thru to next instruction).

RULE-01.
    COMPUTE CLERK = 1.
    GO TO RULE-COMMON-1.
RULE-02.
    COMPUTE CLERK = 2.
RULE-COMMON-1.
    WRITE CLERK-RECORD.
    GO TO READ-IT.

CLOSE-UP SECTION.
CLOSE-UP-1.
    CLOSE CUSTOMER-FILE, CLERK-FILE.
    STOP RUN.
```

Figure 17-8 Host program (including decomposition of Figure 17-7 decision table).

Part V

TABLE TRANSLATION

SECTION 18

Decomposition Algorithms

DECOMPOSITION OF THE DECISION TABLE

Decision tables represent convenient ways to present logical structures of the type commonly seen in data processing situations. But the mechanics of data processing—the program—usually requires logical structures to be expressed in a set of two-way (binary) decisions.

The conversion of decision tables to a set of two-way decisions useable as a program is accomplished by a process known as decomposition.

Decomposition of a decision table is, as of this writing, much in the same boat as the computerized playing of chess: no mathematically proved procedure has been found to determine the best "next move" at any point in a game, except for the "exhaustive" approach of following out every alternative to the final result and picking the one which works best. Decision table decomposition does have one big advantage over chess playing in that there is no opponent; there is no other factor than the "current status" of the table at some given time and the operation of the algorithm (the one "player" involved) to be considered.

CONVERSION OF EXTENDED ENTRY TO LIMITED ENTRY

Because decomposition procedures generally operate on limited entry tables, the presence of extended entry conditions or actions requires a conversion of these entries to limited entry form prior to the actual conversion. Three cases of conversion are discussed: *normal* extended entry conditions, *bounded range* extended entry conditions, and *extended* entry actions.

Conversion of Normal Extended Entry Conditions

Normal extended entry conditions are those in which each rule entry is either identical to or mutually exclusive from each of the other entries, or is a don't-care. Each different entry (except don't-cares) is made into a separate limited entry condition row. Each entry that gives rise to a limited entry row is added onto the stub of the extended entry row to form the stub of the generated limited entry row.

Those entries in the new limited entry row corresponding to the entry in the old extended entry row that contained the stub value is represented by a Y entry. Those that were don't-cares in the original extended entry row remain don't-cares in the new limited entry rows, while all of the remaining entries become * entries. The reason for the choice of the * entry is discussed later under *Horizontal Effects*. The conversion of a normal extended entry to an equivalent set of limited entry condition rows is illustrated in Figure 18-1.

Conversion of Bounded Range Extended Entries

Bounded ranges used as extended entry conditions are converted to limited entry much the same way as were the normal extended entry conditions.

Bounded nonoverlapping ranges are described by *two* condition row entries, one being a greater (or a *greater or equal*) operator, the other a less (or a *less or equal*) operator. Thus a pair of condition rows defines rule entries which, as ranges, are either identical or mutually exclusive to one another (or are don't-cares).

But either of the two rows, taken separately, may not individually appear at first glance to have any mutually exclusive rule entries; for example, greater than 1 in one rule appears not to be mutually exclusive with a greater than 5 in another entry, as witness a value, say, of 7 which meets both criteria. However, if it is known that a particular condition row is a member of a bounded range pair, then it is safe to assume that the entries are in fact either identical, mutually exclusive or don't-cares, just as with normal extended entry condition rows. The only difference is the assignment of the implicit values * and $. In converting normal extended entry conditions to limited entry, all rules corresponding to values mutually exclusive with the stub value received *. In bounded LT (less than operator) condition rows, rules corresponding to values greater than the stub values receive *,

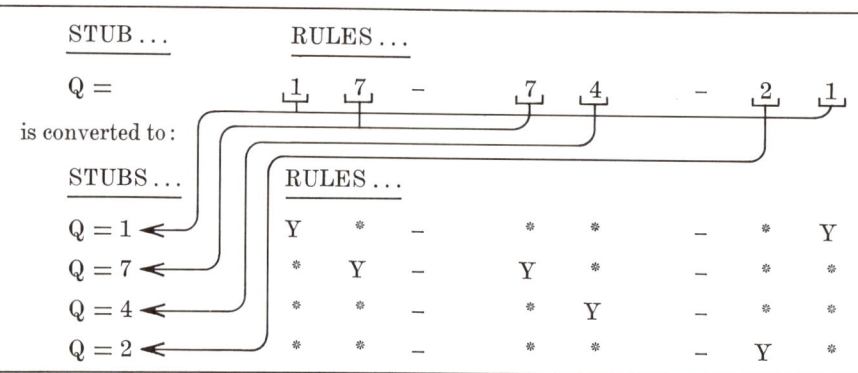

Figure 18-1 Conversion of normal extended entry condition row to limited entry.

CONVERSION OF EXTENDED ENTRY TO LIMITED ENTRY

STUBS...	RULES...				
X GT	1	5	–	5	17
X LT	4	9	–	9	21

Converts to

X > 1	Y	$	–	$	$
X > 5	*	Y	–	Y	$
X > 17	*	*	–	*	Y
X < 4	Y	*	–	*	*
X < 9	$	Y	–	Y	*
X < 21	$	$	–	$	Y

Y *for rules in which stub values appeared.*
– *for rules where "–" appeared.*
* *or $ for other rules depending on the operator (>, <) and relative values of the original rule entries.*

Figure 18-2 Conversion of nonoverlapping bounded range extended entry condition rows to limited entry.

those rules corresponding to values less than the stub value receive a $.

GT (greater than operator) rows reverse the assignment of * and $. Those rows whose operators contain the equal case (GE and LE) involve an added negation of the Y entry into an N entry.

Figure 18-2 serves to illustrate the conversion of two condition rows that make up a bounded nonoverlapping range pair into a total of six limited entry condition rows.

The automatic conversion of bounded range operators *with overlapping ranges* is *not* covered here. Only the nonoverlapping range bound operators are considered in this section.

Extended Entry Action Rows

Action rows are converted to limited entry rows in much the same way as normal extended entry (see above) except that no * entries are used. Those rules in the new limited entry rows that correspond to the stub values receive an X entry, all others a don't-care entry, as shown in Figure 18-3.

Horizontal Effects

It may have been noticed that the * and $ assignments in the above conversions have been made on a "horizontal" basis (on each limited entry condition row individually) rather than "vertically" as the section on *The Condition* suggested that such assignments be made. There is good reason for using this horizontal approach for computer conversion. During the computer conversion of an extended entry condition, we have available a good bit of information about what the other rows will look like and can thus assume that certain condition dependencies will exist in the final, converted table even though they do not necessarily exist in computer core at the point in time when the substitution is made. Figure 18-1 (Normal Extended Entry) and

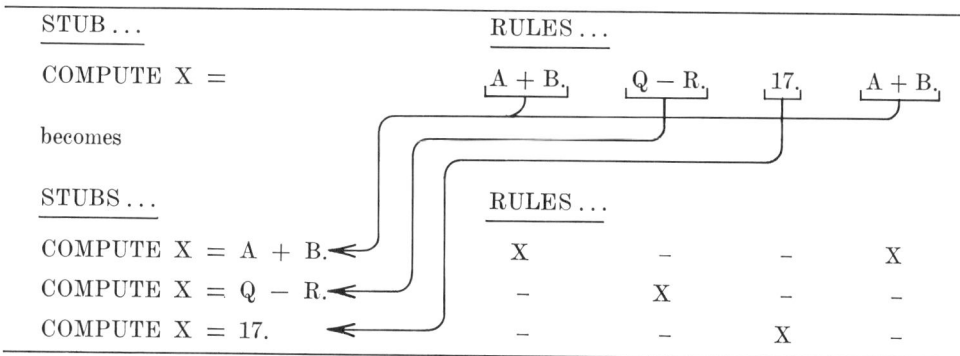

Figure 18-3 Conversion of extended-entry action row to limited entry.

Figure 18-2 (Bounded Range Extended Entry) bear out the fact that horizontal placement, in anticipation of providing the proper vertical requirements, is a sound procedure.

Preprocessors can be designed to allow decision-table users to employ a special symbol to indicate the extremes of a range of values. These can be extremely useful when the extreme limit itself is to be included within a range. Since need for this type of indicator can vary widely, specific rules for its use are not presented here.

THE DECOMPOSITION TREE

After all conversion of extended entry condition and action rows to limited entry (the action rows are no longer of interest from the point of view of decomposition and are dropped at this point) the table is entirely limited entry format.

The process of decomposing the table consists of picking a particular condition row for testing, dividing the table into two parts (or subtables) based on the results of that test, picking another row in each of these two subtables, to split each of those two into two further subtables, and so on until each part represents exactly one rule. This results in a tree-like structure, as illustrated in Figure 18-4. The "limbs" of this tree end whenever a single rule has been isolated. Actually, there may be a few conditions to check before that rule can be selected for execution; that is, the tree can be extended until each rule is selected, rather than just until a one-rule sub-table has been reached.

The Decomposition Algorithm

Many different approaches have been taken to an algorithm[1] for selecting the particular condition on which the table might be split.[2] One particular algorithm for selecting the particular row (called the c_k row) to be used as the basis of division is discussed here. Definitions of the terms used in the algorithm follow.

Dash Count (DC) *for a Row*

Dash count is the number of dashes appearing in that row.

Example: Y - - N - -

The above row's DC = 4.

Delta Count (ΔC) *for a Row*

Delta count is the absolute value of the difference between the number of Y's in a row and the number of N's in that row.

Example: - Y - N N N Y N N
$\Delta C = |2 - 5| = |-3| = 3$

Asterisk/Dollar Count (ADC) *for a Row*

Asterisk/dollar count is the total number of asterisks and dollar signs appearing in that row.

Example: Y $ * $ N * Y $
ADC = 5

Split Count (SC) *for a Row*

For each row eligible to be selected, a one is tallied whenever, for the columns where all the Y's appear in the eligible row, there does not exist a total string of equivalent characters in one of the remaining rows of the table. Similarly, for the columns where all N's appear in the eligible row.

[1] An algorithm is a procedure by which a particular problem might be solved.

[2] See Appendix III, *A Review of Decomposition Techniques.*

Example: Row 1 Y N N Y eligible row
Row 2 Y Y – –
Row 3 N N – –
Row 4 Y Y N N eligible row
Row 5 – – N N
Row 6 Y N N Y eligible row

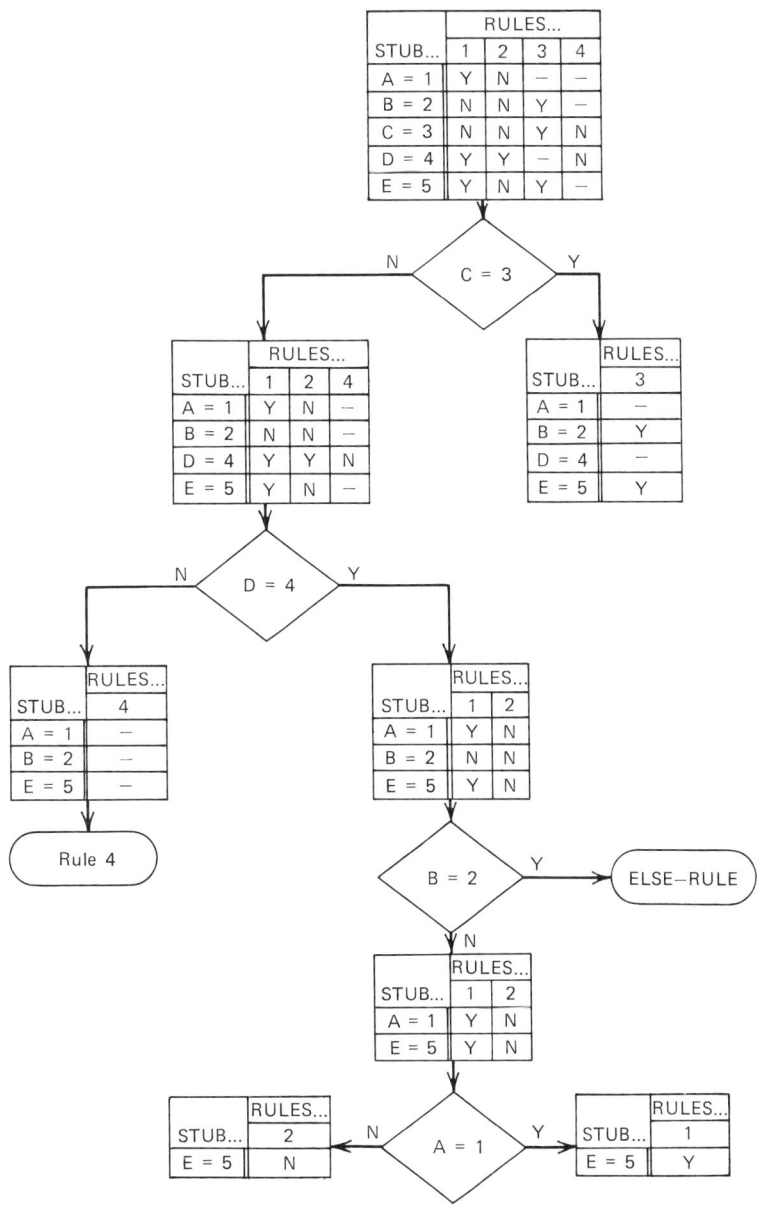

Figure 18-4 Decomposition tree of a decision table.

To illustrate further, this example is rewritten twice—first showing only those columns containing a Y for Row 1 and then showing only those columns containing an N for Row 1:

```
Y X X Y   (base row)
Y X X -   (tally 1)
N X X -   (tally 1)
Y X X N   (tally 1)
- X X N   (tally 1)
Y X X Y   (no tally)
          ─
          4
```

```
X N N X   (base row)
X Y - X   (tally 1)
X N - X   (tally 1)
X Y N X   (tally 1)
X - N X   (tally 1)
X N N X   (no tally)
          ─
          4
```

This illustrates the Row 1 split-count calculation, with the total $= 4 + 4 = 8$. In determining the split count for Row 1, we tally 2 for Row 2, 2 for Row 3, 2 for Row 4, 2 for Row 5 and 0 for Row 6. The SC for Row 1 $= 2 + 2 + 2 + 2 + 0 = 8$. To determine the split count for Row 4, tally 2 for Row 1, 0 for Row 2, 0 for Row 3, 0 for Row 5 and 2 for Row 6. The SC for Row 4 $= 2 + 0 + 0 + 0 + 2 = 4$. Row 6's SC, similarly calculated, is 8.

SAMPLE DECOMPOSITION ALGORITHM TO MINIMIZE NUMBER OF DECISION BRANCHES

Step 1: Select row that has minimum DC. If tie, from among eligible rows that have no $ or * in the row,

Step 2: select the first row that has all Y or all N (ignoring the dashes, if any, that are present). If there is no row that has all Y or all N (ignoring dashes),

Step 3: determine if there are row(s) with one or more * or $. If none of the eligible rows has a * or $, go to step 7, otherwise go to next step.

Step 4: Among the eligible * / $ rows, determine all rows that have at least 2 Y's or 2 N's. If none of these eligible rows has at least 2 Y's or 2 N's, go to Step 7, otherwise go to next step.

Step 5: Among the * / $ rows that have at least 2 Y's or at least 2 N's, select, as C_k, the first row that has 2 or more of its Y's (or N's) in the same column as an eligible row (without * or $) that has at least one Y/N pair in those columns. If there is no such row among the */$ rows, go to next step.

Step 6: Select from among the eligible asterisk/dollar rows the one with minimum ADC. If there is still a tie, select any one.

Step 7: From among the eligible non-asterisk/dollar rows, select the row with maximum Delta Count.

Step 8: If there is a tie in the maximum Delta Count, select (from among the eligible rows) the row with minimum Split Count. If there is still a tie, choose any one of the rows that has the minimum Split Count.

Note: If among the two or more eligible rows there are only *, $ or dash entries and there is more than one rule, contradiction or redundancy exists.

The above algorithm is described in decision-table format in Figures 18-5 and 18-6 and its use is given in Figure 18-7.

DECOMPOSITION ALGORITHM TO MINIMIZE NUMBER OF DECISION BRANCHES

	R1	R2	R3	R4	R5	R6	R7	R8
Only 1 eligible row with minimum DC?	Y	*	*	*	Y	N	*	*
Two or more eligible rows with minimum DC?	*	Y	Y	Y	Y	N	Y	Y
At least 1 eligible non */$ row has all Y's or all N's (not including dashes)	–	Y	N	N	–	–	–	Y
Any eligible row an */$ row?	–	–	N	Y	–	–	N	Y
Eligible non */$ row with maximum Delta	–	$	Y	–	–	–	N	N
Row with minimum DC = C_k	X							
Non */$ row with all Y's or N's (excluding dash) = C_k		X						
Non */$ row with maximum Delta = C_k			X					
Go to Figure 18-6 (*/$ rows)				X				
Impossible case					X		X	
No condition rows in table						X		
Nonallowable check								X

Figure 18-5 Minimization algorithm summarized in decision table format.

	R1	R2	R3	R4	R5	R6	R7	R8	R9	R10	R11
Are there any eligible non */$ rows?	N	Y	Y	Y	Y	Y	N	Y	Y	Y	Y
Eligible row with minimum ADC in */$ row.	Y	-	-	-	-	-	N	-	-	-	-
One or more eligible rows have at least 2 Y's or 2 N's?	-	Y	Y	Y	N	N	-	N	Y	N	Y
Does eligible */$ row RELATE to eligible non */$ row?	-	Y	N	N	-	-	-	-	N	-	N
Only 1 eligible non */$ row with maximum Delta?	-	-	Y	N	Y	N	-	N	N	N	N
Two or more eligible non */$ row with maximum Delta?	-	-	-	Y	-	Y	-	N	N	Y	Y
Eligible non */$ row with minimum Split Count	-	-	-	Y	-	Y	-	-	-	N	N
Any */$ row with minimum ADC $= C_k$	X										
Any */$ row (with 2 or more Y's (or N's)) that relates to a non */$ row $= C_k$		X									
The one non */$ with maximum Delta $= C_k$			X		X						
Any row with minimum Split Count $= C_k$				X		X					
Impossible case							X	X	X	X	X

Note 1: Prior to entering this table all eligible non */$ rows have been checked for all Y's or all N's (*not including dashes*). Hence at this point, all eligible non */$ rows (*if present*) have at least one Y/N pair. Also, there is at least one eligible */$ row.

Note 2: We define an */$ row as being RELATED TO a non */$ row, when any two Y's (*or any two N's*) of the */$ row are in the same columns as a Y/N pair of a non */$ row.

Figure 18-6 Summary of check of */$ rows in decision table format.

DECOMPOSITION ALGORITHM TO MINIMIZE NUMBER OF DECISION BRANCHES

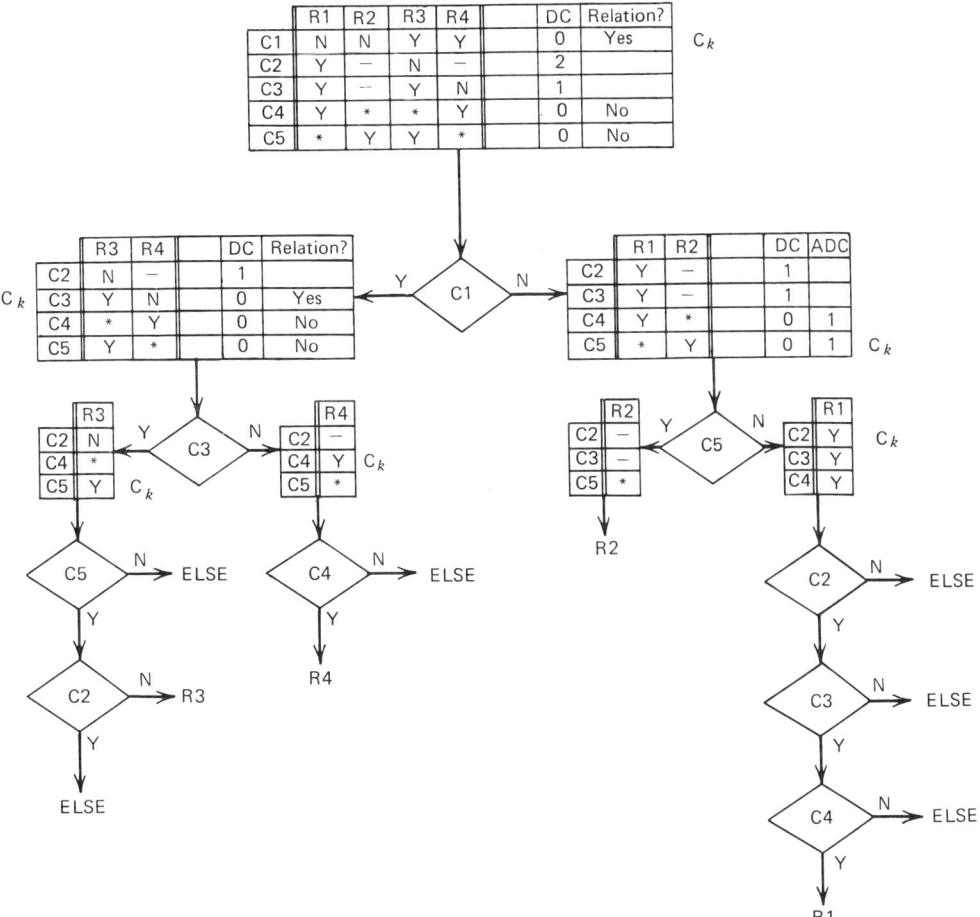

Figure 18-7 Transformation of a decision table to a minimum number of decision boxes by algorithm described in text.

Bibliography

Armerding, G. W., FORTAB: A Decision Table Language for Scientific Computing Applications, Rand Corp., RM-3306-PR Sept. 1962, 39 pp.

Barnard, T. J., "A New Rule Mask Technique for Interpreting Decision Tables." *The Computer Bulletin*, **13**, May 1969, p. 153.

Boerdam, W., *Decision Tables in System Design*, Atlantic-Richfield Co., Los Angeles, Calif., (unpublished), 9 pp.

Boeing, *Decision Tables Training Manual*, The Boeing Company, Aerospace Div., Training Section, Industrial Relations, 1962.

Brown, L. M., "Decision Tables—An Electronic Systems Tool," (date unknown).

Brown, L. M., "Decision Table Experience on a File Maintenance System," *Proc. Decision Tables Symp.*, Sept. 1962, pp. 75–80.

Broyden, C. G., "Solving Nonlinear Simultaneous Equations," *Maths of Comp.*, **19**, 1965, p. 577.

Buckerfield, P. S. T., "A Technique for the Construction and Use of a Generalized Information System," *Proc. IFIP Congress*, Software I, Booklet B, 1968, pp. 61–68.

Calkins, L. W., "Place of Decision Tables and DETAB-X," *Proc. Decision Tables Symp.*, Sept. 1962, pp. 9–12.

Canning, R. G., "Decision Structure Tables," (unpublished).

Canning, R. G., Time to Consider Structure Tables and EDP Design Sessions. *EDP Analyzer*, **1**(4), May 1963.

Canning, R. G., "How to Use Decision Tables." *EDP Analyzer*, **4**(5), May 1966.

Cantrell, N. H., and J. King, F. G. H. King, "Logic Structure Tables." *Comm. ACM*, **4**(6), June 1961, pp. 272–275.

Cantrell, N. H., "Commercial and Engineering Applications of Decision Tables," *Proc. Decision Tables Symp., Sept.* 1962, pp. 55–61.

Chapin, Ned, "A Guide to Decision Table Utilization," *Data Processing Proc. 1966 DPMA CONF*, **IX**, Oct. 1966, pp. 327–329.

Chapin, Ned, "Parsing of Decision Tables." *Comm. ACM*, **10**(8), Aug. 1967, pp. 507–510, 512.

Chapin, Ned, "An Introduction to Decision Tables." *DPMA Quart.*, **3**(3), April 1967, pp. 3–23.

Chapman, A. E. & M. A. Callahan, "A Description of the Basic Algorithm used in the DETAB/65 Preprocessor." *Comm. of ACM*, **10**(7), July 1967, pp. 441–446.

Clarke, D. A., "Structured Table Logic," *Share Secretary Distribution*, SSD(96), Oct. 1962.

CODASYL Systems Group & Joint Users Group of ACM, *Proc. Decision Tables Symp.*, Sept. 1962.

CODASYL Systems Development Group, *Decision Tables Tutorial Using DETAB/X*, 1962.

CODASYL Systems Group, DETAB-X, *Preliminary Specifications for a Decision Tables Structured Language*, 1962.

CODASYL, Decision Table Task Force of Systems Committee, *Draft of Decision Table Standards*, March 1966.

Cuvelette, J. N., "Application des Tables de decision a l'automatisation des etudes de production et des plannings de fabrication." *Automatisme Paris*, (10), 1966, pp. 552–559.

Denolf, H. Decision Tables: "An Annotated Bibliography. *IAG Quarterly*," (1), 1968, pp. 67–82.

Denolf, Henri, *Decision Tables, a Status Quaestionis*, Katholicke Universiteit,

Lenven-Nederlands, Instituut voor Toegepaste Ekonomische Wetenschappen, Sept. 1968.

Devine, D. J., LOBOC, *Logical Business Oriented Coding*, Ins. Co. of North America, Oct. 1962.

Devine, D. J., "Decision Tables as a Basis of a Programming Language." *DPMA*, **7**, 1965, pp. 461–466.

Devine, D. J., *Decision Table Seminar Student Text*, Trilog Associates, 1966.

Dixon, P., "Special Report, Decision Table Symposium," *Standard EDP Repts.*, **1**, Dec. 1962.

Dixon, P., "Decision Tables and their Application." *Computers and Automation*, **13**(4), Apr. 1964, pp. 14–19.

Dolotta, T. A., E. J. McCluskey, Jr., "Encoding of Incompletely Specified Boolean Matrices", *Proc. Western Joint Comp. Conf.*, **17**, 1960, pp. 231–238.

Egler, J. F., "A Procedure for Converting Logic Table Conditions into an Efficient Sequence of Test Instructions." *Comm ACM* **6**(9), Sept. 1963, pp. 510–514.

Ellis, J., *Decision Tables, A User's Guide*, Western Electric Co., June 1967.

Evans, O. Y., *General Information Manual, Advanced Analysis Method for Integrated Electronic Data Processing*, IBM, F20-8047, 1960.

Evans, O. Y., "Advanced Analysis Method for Integrated Electronic Data Processing," *IBM Gen. Info. Man.*, F20-8047, 1961.

Evans, O. Y., *Reference Manual for Decision Tables*, IBM Corp., Sept. 1961.

Evans, O. Y., "A Method for Systematic Documentation—Key to Improved Data Processing Analysis," *Computer Applications*, Macmillan, 1961, pp. 14–34.

Fife, R. C., "Decision Tables, UNIVAC Application Report," Spring Joint Computer Conf. of Systems & Proc. Assn., April 1965.

Fife, R. C., *Decision Tables*, Systems Programming Dept., UNIVAC, 1966.

Fisher, D. L., "Data Documentation and Decision Tables." *Comm. ACM,* **9**, Jan. 1966, pp. 26–31.

Fiske, F. H., "A Case History: Accounting Use of Decision Tables for Definition of Computer Program Logic," (unpublished). One-Day Technical Symp. of L. A. Chapter of ACM, March 1967.

Fletcher, H. R., Seminar on Decision Tables, Bureau of the Census, Sept. 1969.

Gildersleeve, Thomas R., *Decision Tables and Their Practical Application in Data Processing*, (Englewood Cliffs: Prentice-Hall, 1970).

General Electric Co., GE-225 TABSOL Reference Manual and GE-224 TABSOL Application Manual, CPB-147B, June 1962.

Glans, T. B., "Progress in Decision Table Applications," *Ideas for Management*, 1963 Int'l Systems Meeting, Systems and Proc. Assn., 1963.

Glans, T. B., B. Grad, "Tabular Descriptive Language," *IBM Tech. Rept. 245*, Jan. 1962.

Glans, T. B., B. Grad "7080 Decision Table System Preliminary Manual," *IBM Tech. Rept.* 2D1, April 10, 1962.

Gold, V. and C. Bismuth, "Les Tables de Decision Appliquees a la paye." *Automatisme Paris*, **XI**(10), Oct. 1966, pp. 560–565.

Grad, B., "Tabular Form in Decision Logic." *Datamation*, **II**(7), July 1961, pp. 22–26.

Grad, B., "Decision Tables in Systems Design," Digest of Tech. Papers, ACM National Conf., Sept. 1962, pp. 76–77.

Grad, B., "Structure & Concept of Decision Tables," *Proc. Decision Tables Symp.*, Sept. 1962, pp. 19–28.

Grindley, C. B. B., "Systematics—A Non-Programming Language for Designing and Specifying Commercial Systems for Computers." *The Computer Journal*, **9**, 1966, p. 124.

Grindley, C. B. B., "The Use of Decision Tables within Systematics." *The Computer Journal*, **V11**, No. 2 (Aug. 1968), pp. 128–133.

Hawes, M. K., "The Need for Precise Definition," *Proc. Decision Tables Symp.*, Sept. 1962, pp. 13–18, 20–21.

Hawes, M. K., "The Use of Decision Tables for Problem Specification," *Proc. UNIVAC Users Assn.* April 1965, pp. 55–61.

Hirschhorn, E., "Simplification of a Class of Boolean Functions." *J. ACM*, **5**, 1958, pp. 67–75.

Holstein, D., "Decision Tables, A Technique for Minimizing Routine Repetitive Design." *Machine Design*, **34**(18), Aug. 1962, pp. 76–79.

Honeywell, Inc., *An Introduction to Decision Tables—A Programmed Text*, First Edition, October 1969.

Hoyle, F., "The Relative Abundances of Ni58 and Ni60," *Transcript, IBM Symp. on Computers Research & Education IBM (UK)*, 1963.

Hughes, M. L., R. M. Shank, and E. S. Stein, *Decision Tables*, MDI Publications, 1968.

IBM Corp., 1401 Decision Logic Translator H20-0063; 1401 Decision Logic Translator H20-0068; System/360 Decision Logic Translator, H20-0492. IBM Corp. *Decision Tables—Practice Problems and Solutions*, 1963.

IBM Corp. *Decision Tables—A Systems Analysis and Documentation Technique, IBM Corp.*, F20-8102, 1962.

Kavanagh, T. F., "TABSOL—A Fundamental Concept for Systems Oriented Languages," *Eastern Joint Computer Conf.*, **18**, Dec. 1960, pp. 13–15, 117–136.

Kavanagh, T. F., "TABSOL—the Language of Decision Making." *Computers & Automation*, **10**(9), Sept. 1961, pp. 15, 18–22.

Kavanagh, T. F. and M. Allen, The Use of Decision Tables, *Proc. 1963 Conf. of Int'l. DPMA* (D.P. VI), p. 318.

Kavanagh, T. F. and D. T. Schmidt, *Using Decision Structure Tables, Part I: Principles & Preparation, Part II: Manufacturing Applications*, Datamation, Feb. 1964.

King, J. E., *LOGTAB: A Logic Table Technique*, G. E., March 1959.

King, P. J. H., "Conversion of Decision Tables to Computer Programs by Rule Mask Techniques." *Communications, ACM*, **9**(11) Nov. 1966, pp. 796–801.

King, P. J. H., Some Comments on Systematics. *The Computer Journal*, **10**, 1967, p. 116.

King, P. J. H., "Decision Tables." *The Computer Journal* **10**, Aug. 1967, pp. 135–142.

King, P. J. H., "Ambiguity in Limited Entry Decision Tables." *Comm. ACM*, **II**(10), Oct. 1968.

King P. J. H., "The Interpretation of Limited Entry D.T. Format and Relationships Among Conditions." *The Computer Journal*, **12**, 1969, p. 320.

Kirk, H. W., "Use of Decision Tables in Computer Programming," *Comm. ACM*, **8**, Jan. 1965, pp. 41–43.

Klick, D. C., "TABSOL," Preprints of Summaries of papers presented at Nat'l. ACM, Paper 10, B-2, Sept. 1961.

Larsen, R. P., "Data Filtering Applied to Information Storage & Retrieval Applications." *Comm. ACM*, **9**, 1966, p. 785.

Little, J. D. C., K. G. Murty, and D. W. Sweeney, and C. Kareb, "An Algorithm for the Traveling Salesman Problem," *Oper. Res.* II, 1963, pp. 972–989.

Lombardi, L. A., "A General Business-Oriented Language Based on Decision Expressions." *Comm. ACM*, Feb. 1964, pp. 104–111.

Ludwig, H. R., "Simulation with Decision Tables." *Journal of Data Management*, **6**, Jan. 1968, pp. 20–27.

McDaniel, H. (Editor), *Application of Decision Tables*, Auerbach Publishers, Inc., 1970, 276 pp.

McDaniel, H. (Editor), *Decision Table Software*, Auerbach Publishers, Inc., 1970, 70 pp.

McDaniel, H., *An Introduction to Decision Logic Tables*, John Wiley & Sons, 1968, 96 pp.

Montalbano, M., "Tables, Flowcharts and Program Logic." *IBM Systems-Journal*, Sept. 1962, pp. 51–63.

Montalbano, M., Letter to Editor (Egler's Procedure Refuted). *Comm. ACM*, **7**(1) Jan. 1964, p. 1.

Morgan, J. J., "Decision Tables," *Management Services*, Jan. Feb. 1965, pp. 13–18.

Muthukrishnan, C. R., and Rajaraman, V. "On the Conversion of Decision Tables to Computer Programs." *Communications of the ACM*, **V.13** No. 6 (June 1970) pp. 247–251.

Naramore, F., "Application of Decision Tables to Management Information Systems

(Sutherland Co.)," *Proc. Decision Tables Symp.*, Sept. 1962, pp. 63–74.

Nickerson, R. C., "An Engineering Application of Logic Structure Tables." *Comm. ACM,* **4**(11) Nov. 1961, pp. 516–520.

Peel, Roger, "Decision Table Translation." *The Computer Bulletin,* **13**(12) Dec. 1969.

Pollack, S. L., "What is DETAB-X?" *Proc. Decision Tables* Symp. Sept. 1962.

Pollack, S. L., "DETAB-X: An Improved Business-Oriented Computer Language," Rand Corp. Memo RM-3273-PR, Aug. 1962.

Pollack, S. L. and K. R. Wright, "Data Description for DETAB-X," Rand Corp., Memo RM-3010-PR March 1962.

Pollack, S. L., "Analysis of the Decision Rules in Decision Tables," The Rand Corp., Memo RM-3669-PR May 1963.

Pollack, S. L., "How to Build and Analyze Decision Tables," The Rand Corp. memo, P-2829 Nov. 1963.

Pollack, S. L., "CODASYL, COBOL and DETAB-X." *Datamation* **9**(2), Feb. 1963, p. 61.

Pollack, S. L., "The Development and Analysis of Decision Tables," *Ideas for Management,* 1964 Int'l. Systems Meeting, Systems & Procedures Assn., 1964.

Pollack, S. L., "Decision Tables for Systems Design." *Data Processing,* **7**, DPMA, 1965.

Pollack, S. L., "Conversion of Limited Entry Decision Tables to Computer Programs." *Comm. ACM,* **8**(11) Nov. 1965, pp. 677–682.

Pollack, S. L., Letter to Editor Concerning Sprague's Contributions. *Comm. ACM,* **9**(5) May 1966.

Pollack, S. L. and W. J. Harrison, *DETAP Version III User's Guide,* IMI, July 1969.

Pollack, S. L., "Comment on the Conversion of Decision Tables to Computer Programs," *Communications of the ACM,* **V.14** No. (Jan. 1971).

Press, L. I., "Conversion of Decision Tables to Computer Programs." *Comm. ACM,* **8**(6) June 1965, pp. 385–390.

Rabin, J., *Non-Procedural Decision Structures,* (unpublished).

Reinwald, L. T. and R. M. Soland, "Conversion of Limited Entry Decision Tables to Optimal Computer Programs, I: Minimum Average Processing Time." *Jour. ACM,* **13**(3) July 1966, pp. 339–358.

Reinwald, L. T., *An Introduction to TAB40,* Research Analysis Corp., Nov. 1966.

Reinwald, L. T. and R. M. Soland, "Conversion of Limited Entry Decision Tables to Optimal Computer Programs, II: Minimum Storage Requirements." *Jour. ACM,* **14**(4) Oct. 1967, pp. 742–758.

Robinson, Frank, "Processing of Decision Tables in COBOL," *Computer Weekly,* No. 222/223 (December 17/24, 1970).

Roullard, G. P., *The Logic of Switching Circuits,* Exposition University, 1967.

Quine, W. V., "The Problem of Simplifying Truth Functions." *Amer. Math. Mon.,* **59**, 1952, pp. 521–531.

Quine, W. V., "A Way to Simplify Truth Functions." *Amer. Math. Mon.,* **62**, 1955, pp. 627–631.

Schmidt, D. T. and T. F. Kavanagh, "Using Decision Structure Tables." *Datamation,* **10**, Feb.—March 1964.

Shaw, C. J., "Decision Tables—An Annotated Bibliography," *S. D. C.,* TM-2288/000/00, Dec. 1965.

Shober, J. A. H., "Decision Tables for Better Management Systems." *Systems & Procedures Journal,* March—April 1966, pp. 28–32.

Shwayder, K. *Conversion of Limited-Entry Decision Tables to Computer Programs —A Proposed Modification to Pollack's Algorithm,* Comm. ACM, 14(2), Feb. 1971, pp. 69–73.

Slagle, J. R. "An Efficient Algorithm for Finding Certain Minimum Cost Procedures for making Binary Decisions," *J. ACM,* **11**, 1964, pp. 253–264.

Sprague, V. G., Letters to the Editor (On Storage Space of Decision Tables). *Comm. ACM.,* **9**(5) May, 1966, p. 319.

St. Clair, Paul R. Jr., "Decision Tables Clear the Way for Sharp Selection," *Computer Decisions,* **V.12** No. 2 (Feb. 1970) pp. 14–18.

Taylor, H., *Decision Table Technique for Computer Systems,* Hirschfeld Press, 1968.

Veinott, C. G., Letter to Editor (More on Programming Decision Tables). *Comm. ACM,* **9**(7) July 1966, p. 485.

Veinott, C. G., "Programming Decision Tables in FORTRAN, COBOL, OR ALGOL." *Comm. ACM*, **9**(1) Jan. 1966, pp. 31–35.

Verhelst, M., "Procedures for Finding Optimal and Near Optimal Test Sequences for Applying Rule Mask Techniques in Object Programs Derived from Decision Tables." *IAG Quarterly* **1**, 1968, p. 47.

Verhelst, M., *Beslissingtabellen en hun gebruik*, Institute for Applied Economics, University of Louvain, Belgium, 1968.

Verhelst, M., "A Technique for Constructing Decision Tables." *IAG Quarterly Journal*, **2**(1) 1969, p. 27.

Walli, C. R., "Asynchronous Finite-State Controllers for Dynamic Systems," *NAA Space Division*, SID 66–1890, Jan. 1966.

Weizenbaum, J., "Symmetric List Processor." *Comm. ACM*, **6**, 1963, pp. 524–544.

Wilkes, M. V., "Lists and Why They are Useful." *The Computer Journal*, **7**, 1965, p. 278.

Williams, W. K., "Decision Structure Tables." *NAA Bulletin NY*, (9) 1965, pp. 58–62.

Wright, K. R., "Approaches to Decision Table Processors," *Proc. Decision Tables Symp.*, Sept. 1962, pp. 41–44.

APPENDIX I

Proof of Decision Table Theorems

As a prelude to discussing the topics of decision table completeness and contradiction or redundancy of decision rules, we present the supporting theorems and their proofs. For the decision table theory that is developed here, the underlying axiom is that for any set of values for the variables in the conditions of a decision table, one, and only one decision rule in that table can be satisfied by that set of values. Others are free to establish their own decision table axioms and theory. They must be forewarned that relaxation of any part of the theory presented here automatically negates other portions of the theory.

The foundation for decision tables are the AND-Functions and OR-Functions; theorems for each are presented in turn.

AND-FUNCTION THEOREMS

We now offer some basic AND-Function theorems and their proofs.

THEOREM I. *Within a table, two AND-Functions are independent if in at least one condition, one function contains a Y or $, whereas the other contains either an N or * for that condition.*

Proof of Theorem I. Every $ will be considered as included in Y and every * will be considered as included in N. Let B_r and B_s be two AND-Functions in T, where B_r contains as Y in at least one position, say the k_{th} position, and B_s contains an N in that same position, that is,

$$B_r = W_{1r} \cdot W_{2r} \cdot \ldots \cdot Y_k \cdot \ldots \cdot W_n - 1, r \cdot Wnr$$
$$B_s = W_{1s} \cdot W_{2s} \cdot \ldots \cdot N_k \cdot \ldots \cdot W_n - 1, s \cdot Wns$$

The only possible sets of values of the condition variables that can enable $V(B_r)$ to equal 1 are those that result in S's that have a 1 in the k_{th} position. With every one of those sets of values, $V(B_s) = 0$. Therefore B_r and B_s are independent. This proves the first part of the theorem.

Now, suppose that in every position of each of two AND-Functions B_p and B_q there does not exist a Y in one, and an N in the other. We show that B_p and B_q are dependent. Suppose Y exists in the first d positions, N in the next e positions, I in the remaining positions $(n - (d + e))$ of B_p; that is,

146 APPENDIX I

then
$$B_p = Y_1 \cdot Y_2 \cdot \ldots \cdot Y_d \cdot N_{d+1} \cdot \ldots \cdot N_{d+e} \cdot I_{d+e+1} \cdot \ldots \cdot I_n$$

where
$$B_q = I_1 \cdot I_2 \cdot \ldots \cdot I_d \cdot I_{d+1} \cdot \ldots \cdot I_{d+e} \cdot W_{d+e+1} \cdot \ldots \cdot W_n$$

$$W = Y, N, \text{ or } I$$
$$0 \leq d \leq n, \ 0 \leq e \leq n$$

For S = $(\underbrace{1\ 1\ \ldots\ 1}_{d}\ \underbrace{0\ 0\ \ldots\ 0}_{e}\ a_{d+e+1} a_{d+e+2} \ldots a_n$

where
$$a_j = 1 \quad \text{if} \quad W_j = Y$$

and
$$a_j = 0 \quad \text{if} \quad W_j = N;$$
$$j = d+e+1, d+e+2, \ldots, n$$

both $V(B_p) = 1$ and $V(B_q) = 1$ for all possible values of d and e. Hence B_p and B_q are dependent. We can apply the same logic where Y appears in any d positions, N appears in any e positions, and I appears in the remaining $(n - (d+e))$ positions.

Definitions

Recall that a *pure AND-Function* is one that has no I's, i.e. it has exactly n terms each of which is either a Y, an N, an * or a $. In the remainder of this chapter, P will signify a pure AND-Function; for example, for $n = 5$,

$$P = Y_1 \cdot N_2 \cdot N_3 \cdot Y_4 \cdot Y_5$$
is a pure AND-Function.

A *mixed AND-Function* is one that is not pure; that is, it contains one or more I's. It will be signified by M in the remainder of the chapter. Because a mixed AND-Function contains one or more I's; for example,

$$M = N_1 \cdot Y_2 \cdot Y_3 \cdot I_4 \cdot N_5 \cdot N_6;$$

it can be expressed as a combination of pure AND-Functions, all connected by "EXCLUSIVE OR" operators; for example,

$$M = P_1 \oplus P_2 \oplus P_3 \oplus P_4.$$

How a mixed AND-Function containing one or more I's can be expanded into a combination of pure AND-Functions is shown later.

THEOREM II. Within a Table T, each pure AND-Function is independent of every other pure AND-Function.

Proof of Theorem II. Let Z_{im} be a variable that represents Y_i or N_i. Let

1. $P_m = Z_{1m} \cdot Z_{2m} \cdot Z_{3m} \cdot \ldots \cdot Z_{n-1,m} \cdot Z_{nm}$

Since Z represents one of two requirements, $m = 2^n$, and there exists a subtable of T

2. $E = \begin{bmatrix} P_1 \\ P_2 \\ \cdot \\ \cdot \\ \cdot \\ P_{2^n} \end{bmatrix}$

where $P_1, P_2, \ldots, P_{2^n}$ are the only pure AND-Functions of T.

3. Consider all the pairs (P_r, P_s) of Table E $r = 1, 2, \ldots, 2^n$; $s = 1, 2, \ldots, 2^n$; $r \neq s$
4. P_r and P_s differ from each other in at least one position, say the kth position.
5. Either P_r contains Y_k, and P_s contains N_k or, P_r contains N_k, and P_s contains Y_k.

6. Then P_r and P_s are independent (by Theorem I).
7. This is true for all pairs of AND-Functions of E, and Theorem II is proved.

Corollary to Theorem II. Within a Table T, there exists exactly 2^n pure AND-Functions. Each of the remaining $(3^n - 2^n)$ AND-Functions is mixed.

THEOREM III. The form of a mixed AND-Function that contains I in r positions ($1 \leq r < n$) can be expanded into a *canonical form* that consists of 2^r pure AND-Functions each connected by an exclusive "OR" operator ($\oplus$).

Proof of Theorem III. Let Z_{mt} be a variable that represents Y_m or N_m;

$$m = r+1, r+2, \ldots, n$$

Corollary 1 of Theorem III. The canonical form of a mixed AND-Function contains an even number of pure AND-Functions.

Corollary 2 of Theorem III. The canonical form of a mixed AND-Function contains at least two pure AND-Functions.

THEOREM IV. Within a Table T every mixed AND-Function that contains I in r positions ($1 \leq r < n$) is dependent upon each of 2^r pure AND-Functions of T implied by that mixed AND-Function.

Proof of Theorem IV. Let M be a mixed AND-Function that contains I in r positions; $1 \leq r \leq n$. Then, we can expand M to its canonical form.

1. $M = P_1 \oplus P_2 \oplus P_3 \oplus \ldots \oplus P_{2^r}$ where $P_1, P_2, \ldots, P_{2^r}$ are all contained in T (by Theorem III).

Let
1. $M_t = I_1 \cdot I_2 \cdot \ldots \cdot I_{r-1} \cdot I_r \cdot Z_{r+1,\,t} \cdot Z_{r+2,\,t} \cdot \ldots \cdot Z_{nt}$
2. $I = Y + N$
3. $M_t = (Y_1 + N_1) \cdot (Y_2 + N_2) \cdot \ldots \cdot (Y_{r-1} + N_{r-1}) \cdot (Y_r + N_r) \cdot Z_{r+1,\,t} \cdot \ldots \cdot Z_{nt}$

An accepted theorem in Boolean logic is

$$(R + S) \cdot T = R \cdot T + S \cdot T$$

Applying this theorem to the 1st term of step 3, we get

4. $M_t = [Y_1 \cdot (Y_2 + N_2) \cdot \ldots \cdot (Y_r + N_r) \cdot Z_{r+1,\,t} \cdot \ldots \cdot Z_{nt}]$
$\qquad + [N_1 \cdot (Y_2 + N_2) \cdot \ldots \cdot (Y_r + N_r) \cdot Z_{r+1,\,t} \cdot \ldots \cdot Z_{nt}]$

Repeating this expansion $(r - 1)$ times, we get a tree effect where an "AND" operator connects each element in a line to the next and an inclusive "OR" operator connects each line across to the next. This tree effect is shown in Figure A1-1.

6. Since Y_i and N_i appear in r positions, M_t consists of 2^r pure AND-Functions connected by "+."

$$M_t = P_1 + P_2 + P_3 + \ldots + P_{2^r}$$

where $P_1, P_2, \ldots, P_{2^r}$ are all contained in T.

7. $P_1, P_2, \ldots, P_{2^r}$ are each independent of each other (by Theorem II).
8. For every pair of (P_p, P_q) there exists no set of values of the condition variables, such that both $V(P_p) = 1$ and $V(P_q) = 1$.
9. $\qquad M_t = P_1 \oplus P_2 \oplus \ldots \oplus P_{2^r}.$
10. We showed this theorem to be true when I appears in the first r positions. The logic used in this proof can be applied for I appearing in any r positions.
11. The theorem is therefore proved.

148 APPENDIX I

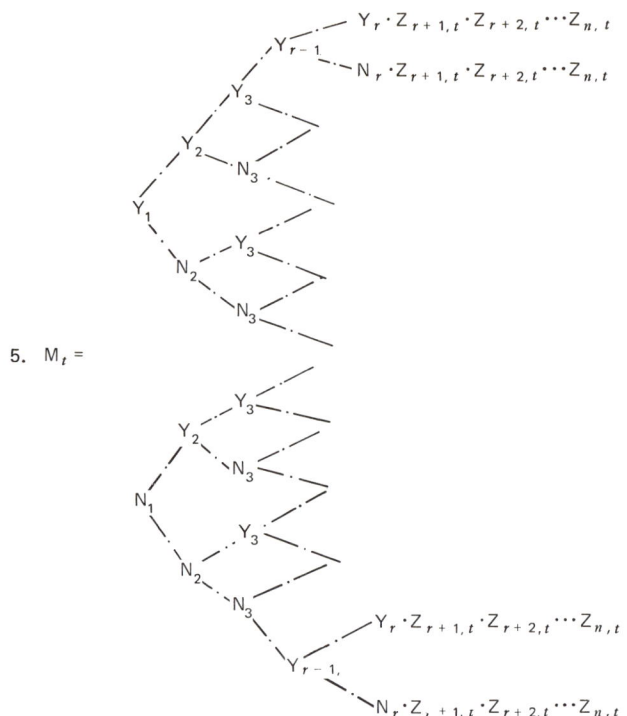

5. $M_t =$

Figure A1-1 Expansion of M_t to a disjunction of pure AND-Functions.

2. When $V(P_1) = 1$, $V(M) = 1$,
∴ M and P_1 are dependent.
When $V(P_2) = 1$, $V(M) = 1$,
∴ M and P_2 are dependent.

. . .
. . .
. . .

When $V(P_{2^r}) = 1$, $V(M) = 1$,
∴ M and P_{2^r} are dependent.

3. M is dependent upon each of 2^r pure AND-Functions of T that are implied by M.

Corollary 1 of Theorem IV. Two mixed AND-Functions are dependent if their canonical forms each contain one or more pure AND-Functions that are common to both; for example, if
$M_1 = P_2 \oplus P_3 \oplus P_7 \oplus P_8 \oplus P_9 \oplus P_{12}$ and
$M_3 = P_1 \oplus P_2 \oplus P_5 \oplus P_6 \oplus P_{11} \oplus P_{13}$,
then M_1 and M_3 are dependent since P_2 is common to both.

Corollary 2 of Theorem IV. A mixed AND-Function is dependent on each of the pure AND-Functions contained in its canonical form.

Corollary 3 of Theorem IV. If a pure AND-Function and a mixed AND-Function are dependent, the canonical form of the mixed function contains the pure function.

Corollary 4 of Theorem IV. If two mixed AND-Functions are dependent, there exists at least one pure AND-Function in their canonical forms that is common to both.

THEOREM V. Table T, based on n conditions, contains one, and only one, set of 2^n independent AND-Functions.

Proof of Theorem V

1. Within a Table T, there exist exactly 2^n pure AND-Functions (by corollary to Theorem II).

2. The 2^n pure AND-Functions are independent of each other (by Theorem II).

Thus, there exists one set of 2^n independent AND-Functions.

3. Now, to show that this is the only set of 2^n independent AND-Functions, we consider all the remaining sets of 2^n AND-Functions that can be formed from the 3^n AND-Functions in T. Denote their domain as R.

4. In each of these sets of R, let there be t mixed AND-Functions and $(2^n - t)$ pure AND-Functions; $1 \leq t \leq 2^n$. 1) For each t, we form all the possible sets of 2^n AND-Functions. 2) The set of 2^n pure AND-Functions is not in R, since $t \neq 0$.

5. Those sets of R, containing one or more pairs of mixed AND-Functions that contain one or more pure AND-Functions common to both, cannot contain 2^n independent AND-Functions (by Corollary 1 to Theorem IV).

6. We therefore look at the Domain Q of R that comprises all those sets of R that contain pure AND-Functions common to the pair.

7. In each of the sets of Q, the canonical form of the t mixed function contains at least 2^t distinct pure functions (by Corollary 2 of Theorem III).

8. In each set of Q, there exist at least $2^n - t + 2^t$ or at least $2^n - t$ pure AND-Functions, where $t \geq 1$.

9. Since there are exactly 2^n pure AND-Functions in T, at least one of them is repeated within each set of Q.

10. If $t = 2^n$, there exists at least one pure AND-Function common to one pair of the 2^n mixed AND-Functions in every set of Q. This violates Step 6 above.

11. Q then contains all sets of t mixed and $(2^n - t)$ pure AND-Functions where $1 \leq t \leq 2^n$; that is, there exists at least one pure AND-Function in every set of Q.

12. In every set of Q, the pair of identical pure AND-Functions (see Step 9 above) exists as one of the functions and as part of a mixed function.

13. Hence, in every set of Q, there exists a pair of dependent AND-Functions (by Corollary 2 of Theorem IV).

14. Therefore, there exists no set in Q that contains 2^n independent AND-Functions.

Complex Decision Rules

THEOREM VI. A complex decision rule that contains I in r positions of its AND-Function is equivalent to 2^r simple decision rules.

Proof of Theorem VI

1. Let $D_k = M_k \rightarrow A_k$ where M_k contains I in r of its positions.

2.
$$M_k = P_1 \oplus P_2 \oplus \ldots \oplus P_{2^r} \text{ (by Theorem III)}$$

3.
$$D_k = (P_1 \oplus P_2 \oplus \ldots \oplus P_{2^r}) \rightarrow A_k.$$

4.
$$D_k = (P_1 \rightarrow A_k) \oplus (P_2 \rightarrow A_k) \oplus \ldots \oplus (P_{2^r} \rightarrow A_k).$$

5. Hence D_k is equivalent to 2^r simple decision rules.

Therefore a complex rule that contains I in r positions of its AND-Function implies 2^r simple decision rules.

OR–FUNCTION THEOREMS

Conversion of an Inclusive OR-Function to an AND-Function

Define $\phi_i = \overline{I_i} = \overline{(Y_i + N_i)} = (N_i \cdot Y_i)$. We denote ϕ as the null requirement. Denote the AND-Function as

$$B_j = W_{1j} \cdot W_{2j} \cdot \ldots \cdot W_{n-1,j} \cdot W_{nj};$$

where W_i represents Y_i, N_i, or I_i; $j = 3^n$. Denote the OR-Function as

$$E_j = U_{1j} + \ldots + U_{n-1,j} + U_{nj},$$

where U_i represents Y_i, N_i, or ϕ_i, $j = 3^n$. $*_i$ is assumed included in N_i, $\$_i$ included in Y_i. Since all our theorems have been postulated for AND-Functions, we now explore the relations between AND-Functions and OR-Functions.

An OR-Function can be converted to an AND-Function with the following procedure, on the basis of the definition of the inclusive "OR" operator.

1. Suppose
$$E_j = U_{1j} + U_{2j} + \ldots + U_{n-1,j} + U_{nj}; U_i \text{ represents } Y_i, N_i, \text{ or } \phi_i.$$

2. If $U_{1j} \neq \phi$, $B_1 = U_{1j} \cdot I_2 \cdot I_3 \cdot \ldots \cdot I_n$
   If $U_{1j} = \phi$, $B_1 = 0$.

3. If $U_{2j} \neq \phi$, $B_2 = \overline{U_{1j}}$
   $\cdot U_{2j} \cdot I_3 \cdot I_4 \cdot \ldots \cdot I_n$
   If $U_{2j} = \phi$, $B_2 = 0$.

4. If $U_{3j} \neq \phi$, $B_3 = \overline{U_{1j}}$
   $\cdot \overline{U_{2j}} \cdot U_{3j} \cdot I_4 \cdot I_5 \cdot \ldots \cdot I_n$
   If $U_{3j} = \phi$, $B_3 = 0$.

   .   .   .   .
   .   .   .   .
   .   .   .   .

5. If $U_{nj} \neq \phi$, $B_n = \overline{U_{1j}}$
   $\cdot \overline{U_{2j}} \cdot \overline{U_{3j}} \cdot \ldots \cdot \overline{U_{n-1,k}} \cdot U^n{}_j$
   If $U_{nj} = \phi$, $B_n = 0$.
   Let G denote E in converted form

6. Let $G = B_1 + B_2 + B_3 + \ldots + B_{n-1} + B_n$
   Where $B_i = 0$ if $U_{ij} = \phi$
   $B_i = \overline{U_{1j}} \cdot \overline{U_{2j}} \cdot \overline{U_{3j}}$
   $\cdot \ldots \cdot U_{i-1,j} \cdot U_{1j} \cdot I_{i+1,j}$
   $\cdot \ldots \cdot I_n$ if $U_{ij} \neq \phi$

Although we started with U_{1j} and continued with U_{2j} then U_{3j}, etc., to convert E to G, we could have started with any U, and taken away any of the remaining U's, etc., to convert E to G. They are, in fact, equivalent; that is, they have the same truth table for all possible values of the condition variables. If each of the B's that contain I's are converted to their canonical forms, the two G's will be found to be identical. This is because both G's were derived on the basis of the definition of the inclusive "OR" operator; for example, let $E = Y_1 + N_2 + N_3$. If we use Y_1, then N_2, then N_3,

$$G_1 = (Y_1 \cdot I_2 \cdot I_3) \oplus (N_1 \cdot N_2 \cdot I_3)$$
$$\oplus (N_1 \cdot Y_2 \cdot N_3)$$

$G_1 = Y_1 \cdot Y_2 \cdot N_3$ 1.
$\oplus Y_1 \cdot Y_2 \cdot Y_3$ 2.
$\oplus Y_1 \cdot N_2 \cdot N_3$ 3.
$\oplus Y_1 \cdot N_2 \cdot Y_3$ 4.
$\oplus N_1 \cdot N_2 \cdot Y_3$ 5.
$\oplus N_1 \cdot N_2 \cdot N_3$ 6.
$\oplus N_1 \cdot Y_2 \cdot N_3$ 7.

If we use N_2 then N_3 then Y_1,

$G_2 = Y_1 \cdot N_2 \cdot Y_3$ 4.
$\oplus N_1 \cdot N_2 \cdot Y_3$ 5.
$\oplus Y_1 \cdot N_2 \cdot N_3$ 3.
$\oplus N_1 \cdot N_2 \cdot N_3$ 6.
$\oplus Y_1 \cdot Y_2 \cdot N_3$ 1.
$\oplus N_1 \cdot Y_2 \cdot N_3$ 7.
$\oplus Y_1 \cdot Y_2 \cdot Y_3$ 2.

Notice that G_1 and G_2 contain the same B's. Hence $G_1 = G_2$.

Example of Conversion

Let $E = Y_1 + N_2 + \phi_3 + \phi_4 + Y_5$,
then $G = (Y_1 \cdot I_2 \cdot I_3 \cdot I_4 \cdot I_5)$
$\oplus (\overline{Y}_1 \cdot N_2 \cdot I_3 \cdot I_4 \cdot I_5)$
$\oplus (\overline{Y}_1 \cdot \overline{N}_2 \cdot \overline{\phi}_3 \cdot \overline{\phi}_4 \cdot Y_5)$,
$G = (Y_1 \cdot I_2 \cdot I_3 \cdot I_4 \cdot I_5)$
$\oplus (N_1 \cdot N_2 \cdot I_3 \cdot I_4 \cdot I_5)$
$\oplus (N_1 \cdot Y_2 \cdot I_3 \cdot I_4 \cdot Y_5)$.

The reader will find Theorems VI', IV' and I' presented out of sequence so that the OR-Function discussion can parallel the AND-Function argument.

Consequences of Conversion

Every OR-Function can be converted (is equivalent) to the exclusive union of one or more independent AND-Functions. The truth table of this converted form is the same as the original function for all possible values of the condition variables.

To illustrate, consider

$$E = U_1 + U_2 + U_3 + \ldots + U_{n-1} + U_n$$

that converts to

$$G = B_1 \oplus B_2 \oplus B_3 \oplus \ldots \oplus B_{n-1} \oplus B_n.$$

If one or more of the requirements, that is, either Y or N is satisfied, $V(E) = 1$. For the same values of the condition variables, there exists one, and only one, B for which $V(B) = 1$. [This is true because each B that contains more than one U (that is either Y or N) contains one original U and the remaining U terms are negations of the original U's. The non-negated U is different for each B.] Since $V(B) = 1$, $V(G) = 1$.

$V(E) = 0$ if, and only if, the negation of every U occurs. If the negation of every U occurs $V(G) = 0$, since each B contains one non-negated U (that is either Y or N) and all U (that are either Y or N) are contained in G. Further, $V(G) = 0$ only if $V(B) = 0$ for every B. This can occur only if the negation of every U occurs.

E and G, therefore, have the same truth table for all possible values of the condition variables.

THEOREM VI'. An OR-Function that contains ϕ in r ($0 \leq r < n$) positions is equivalent to $(2^n - 2^r)$ distinct pure AND-Functions.

Proof of Theorem VI'

1. Consider an OR-Function that contains ϕ in the last r positions, and Z_i (which represents Y_i or N_i) in the first $(n - r)$ positions:

$$E = Z_1 + Z_2 + \ldots + Z_{n-r} + \phi_n.$$

2. E can be expanded to:

$$G = (Z_1 \cdot I_2 \cdot \ldots \cdot I_n) \oplus (\overline{Z}_1 \cdot Z_2 \cdot I_3 \cdot \ldots \cdot I_n)$$
$$\oplus \ldots \oplus (\overline{Z}_1 \cdot \overline{Z}_2 \cdot \ldots \cdot \overline{Z}_{n-r-1}$$
$$\cdot Z_{n-r} \cdot I_{n-r+1} \cdot \ldots \cdot I_n).$$

3. The first term contains $(n - 1)$ I's. It is equivalent to 2^{n-1} pure AND-Functions. The second term contains $(n - 2)$ I's. It is equivalent to 2^{n-2} pure AND-Functions. The $(n - r)$th term contains r I's. It is equivalent to 2^r pure AND-Functions.

4. G is equivalent to $2^r + 2^{r-1} + \ldots + 2^{n-2} + 2^{n-1}$ distinct pure AND-Functions.

5. G is equivalent to $(2^n - 2^r)$ distinct pure AND-Functions.

6. This same reasoning could have been used if I appeared in any r positions. The Theorem is therefore proved.

An AND-Function and an OR-Function are dependent if there exists at least one set of values of the condition variables such that both assume a truth value 1. Otherwise, they are independent; for example, suppose

$$B = Y_1 \cdot N_2 \cdot N_3, \text{ and } E = N_1 + Y_2 + Y_3.$$

Then there exists no S such that $V(B) = 1$ and $V(E) = 1$ since

$$E = N_1 + Y_2 + Y_3 = (\overline{Y_1 \cdot N_2 \cdot N_3}) = \overline{B}.$$

Therefore, B and E are independent.

THEOREM IV'. Every OR-Function that contains ϕ in r positions ($0 \leq r < n$) is dependent on $(2^n - 2^r)$ distinct pure AND-Functions.

Proof of Theorem IV'

1. Let $E = \phi_1 + \phi_2 + \ldots$
$$+ \phi_r + U_{r+1} + \ldots + U_n.$$

2. $E = P_1 \oplus P_2 \oplus \ldots \oplus P_{(2^n - 2^r)}$
where $P_1, P_2, \ldots, P_{(2^n - 2^r)}$ are distinct pure AND-Functions.

3. When $V(P_1) = 1$, $V(E) = 1$
When $V(P_2) = 1$, $V(E) = 1$

.
.
.

When $V(P_{2^n - 2^r}) = 1$, $V(E) = 1$.

4. Hence E is dependent on $2^n - 2^r$ distinct pure AND-Functions.

5. Although ϕ was in the first r positions, the logic of this proof applies no matter in what r positions they appear.

6. The theorem is therefore proved.

THEOREM I'. Two OR-Functions are dependent if, in at least one position, there exists a Y in both, or an N in both. Otherwise they are independent.

Proof of Theorem I'

1. Assume there exists a Y in a particular position of two OR-Functions, C_1 and C_2.
2. A transaction having condition variable values that satisfy the condition in that position of C_1 also satisfies the condition of C_2.
3. $V(C_1) = 1$, $V(C_2) = 1$ for a particular transaction.
4. Hence C_1 and C_2 are dependent.
5. The same proof holds for an N in a particular position of two OR-Functions. Hence C_1 and C_2 are dependent.
6. Assume there exists no position of C_1 and C_2 such that both contain a Y or both contain an N; that is, each position either contains a Y and N, a ϕ and Y, ϕ and N, or ϕ and ϕ.
7. Any transaction that has a condition value in the position where the ϕ appears contributes nothing to make the truth value 0 or 1. Hence those positions that contain ϕ can be ignored.
8. For those cases where one OR-Function contains a Y and the other contains an N, both functions cannot have a truth value 1 for any transaction. Hence C_1 and C_2 are independent and the theorem is proved.

THEOREM I''. An AND-Function and an OR-Function are dependent if, in at least one position, there exists a Y in both, or an N in both, or there exists an I in the AND-Function and a Y or N in the OR-Function. Otherwise they are independent.

Examples

1.
$$B = Y_1 \cdot N_2 \cdot Y_3 \cdot I_4$$
$$E = N_1 + Y_2 + N_3 + \phi_4$$
B and E are independent.

2.
$$B = Y_1 \cdot N_2 \cdot Y_3 \cdot N_4$$
$$E = N_1 + N_2 + N_3 + N_4$$
B and E are dependent.

3.
$$B = Y_1 \cdot N_2 \cdot I_3 \cdot N_4$$
$$E = N_1 + Y_2 + Y_3 + \phi_4$$
B and E are dependent.

Proof of Theorem I''

1. Let $B = Y_1 \cdot W_2 \cdot W_3 \cdot \ldots \cdot W_n$
or
$$B = I_1 \cdot W_2 \cdot W_3 \cdot \ldots \cdot W_n$$
where
$$W_i = Y_i, N_i, \text{ or } I_i$$
and
$$E = Y_1 + U_2 + U_3 + \ldots + U_n$$
where
$$U_i = Y_i, N_i, \text{ or } \phi_i$$
$V(B) = 1$ for $S = (1\, a_2\, a_3 \ldots)$;
$\quad a_i \to [V(W_i) = 1]$ for $i = 2, 3, \ldots$
$V(E) = 1$ for $S = (1\, a_2\, a_3 \ldots)$.
Then B and E are dependent.

2. Similarly for $B = N_1 \cdot W_2 \cdot W_3 \cdot \ldots \cdot W_n$
or
$$B = I_1 \cdot W_2 \cdot W_3 \cdot \ldots \cdot W_n$$
and
$$E = N_1 + U_2 + U_3 + \ldots + U_n$$
$V(B) = 1$ for $S = (0\, a_2\, a_3 \ldots)$;
$\quad a_i \to [V(W_i) = 1]$ for $i = 2, 3, \ldots$
$V(E) = 1$ for $S = (0\, a_2\, a_3 \ldots)$
and B and E are dependent.

3. Regardless where the pair of Y's or the pair of N's or the I in the AND-Function and the Y or N in the OR-Function appear, an S exists with the 1 or 0 in the corresponding position such that $V(B) = 1$. And for that same S, $V(E) = 1$. This proves the first part of the theorem.

4. Suppose there exists no pair of Y's or no pair of N's or no I in the AND-Function with a Y or N in the corresponding position of the OR-Function in any position of B and E.

5. In converting E to AND-Functions, the ϕ contributes no AND-Function so that those positions containing ϕ's can be disregarded.

6. From those positions that contain Y or N, the conversion will yield AND-Functions each of which either

OR-FUNCTION THEOREMS

(a) contains at least one Y in the same position in which B contains an N, or
(b) contains at least one N in the same position in which B contains a Y.

7. Hence, if $V(B) = 1$, $V(E) = 0$, since each of the AND-Functions of the converted form has a truth value 0.

8. Therefore B and E are independent. Q.E.D.

APPENDIX II

An Introduction to COBOL

The purpose of this Appendix is to provide sufficient information about COBOL to illuminate the examples in the text. Those who have some prior knowledge of the language will find nothing new here.

More detailed information on the subject of COBOL is available from several sources. Each computer manufacturer who offers COBOL as one of his languages provides one or more manuals describing his particular implementation (usually a reference manual and perhaps a programmed instruction text or a programmer's guide). The current state of COBOL is documented in the CODASYL COBOL *Journal of Development*,[1] and the American National Standard COBOL in *USA Standard COBOL* (X3.23—1968).[2] Finally, there are a number of textbooks completely or partially devoted to COBOL.

A COBOL source program consists of four Divisions: Identification, Environment, Data, and Procedure. Each Division provides information to the COBOL Compiler, a program that translates the COBOL source program into an object program in the computer's language. The examples used in this book are all procedures, and would appear in the Procedure Division as would the decision tables in which they are found. Thus we will concentrate most heavily on this division.

An example of a statement from a hypothetical Procedure Division is shown below.

    MOVE ITEM-1 TO ITEM-2

This statement and every other Procedure Division statement begins with a COBOL verb—MOVE, in this instance. A verb operates on one or more operands—names representing data items, files, or other Procedure Division statements. The operands in this example are ITEM-1 and ITEM-2 and because the MOVE verb operates on data items, these operands must be defined in the Data Division of the COBOL program. The word TO is part of the format of the MOVE verb and serves to separate the "from" operand and the "to" operand. The statement tells the COBOL compiler to generate machine code to

MOVE (whatever is at) ITEM-1 TO
         (the location of) ITEM-2.

---

[1] Available from the U.S. Printing Office as NBS Handbook #106.

[2] Available from the American National Standards Institute.

COBOL uses a notation to describe the way each of its verbs must be written by the programmer. The MOVE verb would be shown in a COBOL Manual as

<u>MOVE</u> operand-1 <u>TO</u> operand-2 . . .

The words in capital letters (MOVE, TO) are COBOL words and must be spelled as shown. The underline means that the word must always be written when the verb is used. Some verb formats contain COBOL words that are not underlined. These "noise" words can be written or not because they appear in the format simply to improve the readability of the source program.

The words in lower case (operand-1, operand-2) represent programmer defined words. The rules associated with each verb tell what kind of COBOL entities these words must represent. As we mentioned, the rules for MOVE require that both operands be the names of data items. The ellipses (. . .) following "operand-2" indicate that the programmer can write more than one operand-2 in a MOVE statement if he wants to move the data located at operand-1 to more than one place.

Operands that represent data must be defined in the Data Division of the COBOL program. To define data, the programmer writes a series of clauses that tell the COBOL compiler that the characters that will appear in the item when the program runs will be alphabetic (letters), numeric (numbers), or alphanumeric (both); how many characters will be present; and, if the item represents a number, how many positions to the right of the decimal it has. Each data item is given a "level" number that indicates the structural relationship the item has to other items in the same structure.

Suppose we want to define data that contains an address and looks graphically like this:

ADDRESS			
STREET	CITY	STATE	ZIP
NUMBER NAME			

The COBOL definition of this data might appear as follows:

```
01  ADDRESS.
    02  STREET.
            03  ST-NR PICTURE 99999.
            03  ST-NM PICTURE X(20).
    02  CITY        PICTURE A(15).
    02  STATE       PICTURE A(12).
    02  ZIP-CODE    PICTURE 99999.
```

The name of the level 01 entry is called the "record-name," while the entries at levels 02 and 03 are called subordinate entries. The level 03 entries and the last three level 02 entries are called "elementary items" because they have no subordinate entries.

The PICTURE clause enables the programmer to indicate the class (A = letters, 9 = numbers, X = either or both) of an item, and its size (the number of A's, X's, or 9's). Notice that only those items that are at the lowest levels in the structure are defined by "PICTUREs." The characteristics of the items above them in the structure are deduced by the compiler. For example, the length of STREET is 25 characters (5 for ST-NR plus 20 for ST-NM).

Any of the names in the above data definition could appear in the following MOVE statement.

MOVE ST-NR TO ANOTHER-NUMBER
MOVE ZIP-CODE TO ZIP-1, ZIP-2
MOVE ADDRESS TO ADDRESS-SAVE

In all of these statements, the number of characters moved, the first and last characters moved and any conversion or editing depends on the characteristics of the receiving and sending data items.

Another kind of data item, known as a "literal," can be used where a value or constant is needed. For example:

MOVE 'ABC' TO FIELD-1
MOVE 9056 TO CODE-2

These MOVE statements use literals as sending fields rather than named data fields. The literal 'ABC' represents the three letters ABC while the literal 9056 represents the number nine

156   APPENDIX II

thousand fifty six. Some literals are used so often that COBOL gives them fixed names: SPACE(S), representing one or more blank characters and ZERO(S) representing 0000. . . are the two most frequently used.

Certain procedural statements use the names of files as operands; files are collections of records (level 01 entries and their subordinates) that reside on external storage devices such as tapes, drums, or disks. Files are defined in the Data Division, assigned to devices in the Environment Division and accessed through these procedural statements:

$$\underline{\text{OPEN}} \left\{ \begin{array}{c} \underline{\text{INPUT}} \\ \underline{\text{OUTPUT}} \end{array} \right\} \text{file-name} \ldots$$

$\underline{\text{READ}}$ file-name   AT $\underline{\text{END}}$ procedure-statement

$\underline{\text{WRITE}}$ record-name

$\underline{\text{CLOSE}}$ file-name . . .

The OPEN statement must be the first statement addressed to a file during a run. It causes the device to which the file is assigned to be readied for subsequent reading (INPUT) or writing (OUTPUT). Note that the form $\left\{ \begin{array}{c} \underline{\text{INPUT}} \\ \underline{\text{OUTPUT}} \end{array} \right\}$ indicates that the programmer must choose one word or the other and write in that position; for example, OPEN INPUT file-name.

The READ statement causes the next available record from the file to appear in the level 01 area associated with the file. When the file has been emptied, the next READ statement executed will cause the "procedure-statement" to be executed. Note that AT is a "noise" word that may or may not be written at the programmer's discretion.

WRITE causes the data currently in the level 01 named "record-name" to be placed next in the file. It is important to remember that the WRITE statement addresses a record-name, while the READ statement addresses a file-name. CLOSE terminates the processing of the file and deactivates the device.

One or more procedural statements can be given a name (called a procedure-name) that itself can be referred to by other procedures. This facility enables the programmer to vary the logical path that his program follows. The statements that use procedure-names as operands are the following:

$\underline{\text{GO TO}}$ procedure-name

$\underline{\text{PERFORM}}$ procedure-name.

GO TO causes the procedural statement named "procedure-name" to be executed next in sequence. PERFORM is the same as a GO TO followed by a "return"; that is, the instruction following the PERFORM statement will be executed after the last statement in "procedure-name." The following series of procedural statements, for example, will cause this execution sequence: 1, 3, 4, 2, 5, 6.

```
1          PERFORM PARAGRAPH-1.
2          GO TO PARAGRAPH-2.
       PARAGRAPH-1.
3          MOVE FIELD-1 TO FIELD-2.
4          MOVE FIELD-3 TO FIELD-4.
       PARAGRAPH-2.
5          READ FILE-2 AT END MOVE 1 TO SWITCH.
       PARAGRAPH-3.
6          WRITE RECORD-1.
```

So far we have been looking at statements that specify actions to be done—statements called "imperative" statements in COBOL. These are the statements that appear in the Action portion of a decision table. The statement that appears in the Condition section of a table is a portion of the IF statement. The full COBOL IF statement looks like this:

$$\underline{\text{IF}} \text{ condition} \begin{Bmatrix} \underline{\text{NEXT SENTENCE}} \\ \text{procedural-statement-1} \end{Bmatrix}$$

$$[\underline{\text{ELSE}} \text{ procedural-statement-2}]$$

When coding a decision table condition entry, the programmer uses only the portion shown as "condition" in this format. The decision table processor supplies both the word IF and the procedural-statements when it decomposes the table.

COBOL has four forms of conditions, but all perform the same basic function—stating a relationship that can be evaluated as being true or false whenever it is executed at run time. The most commonly used condition is the "relation" condition, stated:

$$\text{operand-1 IS } [\underline{\text{NOT}}] \begin{Bmatrix} \underline{\text{GREATER}} \text{ THAN} \\ \underline{\text{LESS}} \text{ THAN} \\ \underline{\text{EQUAL}} \text{ TO} \end{Bmatrix} \text{ operand-2}$$

The programmer can substitute the symbols > < = for the English equivalents GREATER, LESS, EQUAL TO if he wishes.

Two kinds of conditions, "class" tests and "sign" tests, state a relationship between a single operand and a definition. The class condition is stated:

$$\text{operand IS } [\underline{\text{NOT}}] \begin{Bmatrix} \underline{\text{ALPHABETIC}} \\ \underline{\text{NUMERIC}} \end{Bmatrix}$$

This relationship asks whether the current run-time value of the operand is composed of a string of characters that is all alphabetic or all numeric, *not* whether the operand is defined with a PICTURE containing all A's (ALPHABETIC) or all 9's (NUMERIC).

The sign condition tests the run-time sign of an item and is written:

$$\text{operand IS } [\underline{\text{NOT}}] \begin{Bmatrix} \underline{\text{POSITIVE}} \\ \underline{\text{ZERO}} \\ \underline{\text{NEGATIVE}} \end{Bmatrix}$$

The final type of condition involves the use of a data item called a "condition-name."

```
05  CONDITIONAL-VARIABLE      PICTURE X.
    88  CONDITION-NAME-1      VALUE 'A'.
    88  CONDITION-NAME-2      VALUE 'B'.
```

In this illustration, CONDITION-NAME-1 and CONDITION-NAME-2 are called condition-names. Their purpose is to give a name to the VALUE or "condition" that follows them, in order to permit a shortened version of a relation-condition to be used in the Procedure Division. The condition-name condition is written simply:

$$[\underline{\text{NOT}}] \quad \text{condition-name}$$

and has the same meaning as if the programmer had written:

$$\text{conditional-variable IS } [\underline{\text{NOT}}] = \text{VALUE of condition-name.}$$

158 APPENDIX II

Or, using the names from the illustration, the condition-name condition

CONDITION-NAME-2

means the same as

CONDITIONAL-VARIABLE = 'B'.

We can now see how the IF statement operates. The statement

IF ITEM = 'C' MOVE 1 TO FIELD
ELSE MOVE 2 TO FIELD

causes execution to test the current value of ITEM. If it contains a C, then the condition is true and the procedural-statement immediately following it is executed; in this case MOVE 1 TO FIELD. If ITEM contains something other than a C, then the statement following ELSE will be executed; in this case MOVE 2 TO FIELD.

The condition portion of the IF statement can be compounded by the use of the logical connectives AND and OR; for example,

IF ITEM-1 = 'A' AND ITEM-2 > 4
GO TO NEW-STEP.

In this statement, both simple conditions must be true for control to pass to NEW-STEP. In the statement below, transfer of control occurs as soon as the first true simple condition is encountered:

IF ITEM-3 < 06 OR ITEM-4 > 123
GO TO STEP-6.

Shown below are the formats for all of the procedural statements that appear in this book. The statements are grouped by class and, in some cases, the format has been simplified from the full COBOL form for ease of comprehension.

INPUT-OUTPUT STATEMENTS

OPEN $\left\{ \dfrac{\text{INPUT}}{\text{OUTPUT}} \right\}$ file-name . . .

READ file-name AT END procedural-statement

WRITE record-name

CLOSE file-name . . .

DATA MANIPULATION STATEMENTS

MOVE operand-1 TO operand-2

ADD operand-1 . . . $\left\{ \dfrac{\text{TO}}{\text{GIVING}} \right\}$ operand-2

SUBTRACT operand-1 . . . FROM operand-2
[GIVING operand-3]

MULTIPLY operand-1 BY operand-2 [GIVING operand-3]

DIVIDE operand-1 INTO operand-2 [GIVING operand-3]

COMPUTE operand-1 = arithmetic formula[3]

PROGRAM SEQUENCE CONTROL STATEMENTS

PERFORM procedure-name

GO TO procedure-name

CONDITION-TESTING STATEMENT

IF condition-1 $\begin{Bmatrix} \underline{\text{NEXT SENTENCE}} \\ \text{procedural-statement-1} \end{Bmatrix}$

[ELSE procedural-statement-2]

where: condition is a relation condition, sign condition, class condition, condition-name condition, or a compound condition using AND or OR.

---

[3] An arithmetic formula is a series of operands connected by arithmetic operators; for example ITEM-1 * (ITEM-2 + ITEM-3).

APPENDIX III

Decision Table Translation Algorithms*

Decision tables have proved invaluable as a communications tool in the difficult task of job definition. They provide an easy-to-follow, complete functional description for any person needing information about a job, whether his information need is at the level of a broad summary overview or at the detailed level required by a programmer.

Taken one step farther, decision tables can also be used directly as a tool for communicating job definitions to a computer. For this additional step, an orderly procedure or "algorithm" is needed to translate the tabular structure of the decision table into an efficient sequence of machine-executable instructions. Efficiency is, of course, a relative term and can only be measured with respect to a specifically selected criterion. Several criteria for decision table translation efficiency have been proposed; among them, the time and effort required to make the actual translation, the amount of core storage used by the resultant program and the time required to execute the resultant program. The attempt to optimize the amount of core used is usually called "core storage minimization" while the attempt to optimize the time required for execution is usually called "run time minimization." The two different goals will usually produce different programs from the same table. Consequently, the algorithm selected to translate the decision table depends to a large extent on the efficiency criterion selected.

Once an algorithm has been selected, it can either be used as a hand-coding technique or built into a preprocessor or compiler for automatic translation of tables.

There are two major ways to translate a decision table into a machine-executable sequence of instructions which isolates a unique rule.

1. The first technique is called "scanning" or, at its more sophisticated levels, the "rule mask" technique. It involves testing each transaction against all pertinent conditions in a single rule and scanning across the rules until one is found in which all the conditions are satisfied. The solution is said to be in that the rule and

* This appendix was written by Marjorie Wiggins of Information Management Incorporated.

consequently the actions associated with it are executed. Scanning is a *rule oriented* technique that can handle tables with apparent ambiguities.

2. The second technique is called "condition testing" or the "network technique." This method tests one condition at a time and requires the rules in the decision table to be unambiguous (i.e., one transaction cannot satisfy more than one rule). The network technique takes advantage of this requirement and attempts to rearrange the conditions into an optimal testing network for the isolation of the *unique* rule satisfied by each transaction entry. It is primarily *condition oriented*.

Before dealing with the specifics of algorithms, it is necessary to state what is *not* considered here. Initialization statements and rules, because they must be processed first, are not considered. Extended entry tables are not considered because they can be converted into limited entry tables and are therefore just a special case requiring an extra conversion step. The action portion of a table is not normally considered because the algorithms are designed to isolate a unique rule which, in turn, defines an action set.

SCANNING AND RULE MASK TECHNIQUES

The Straight Scan Technique[1]

To translate a table into a sequence of machine-executable instructions, test the transaction entry against the first pertinent condition of the first rule. If the first condition is satisfied, test the transaction entry against the second pertinent condition of the first rule, otherwise, test the transaction entry against the first pertinent condition of the second rule. Continue until an entire rule is satisfied by the transaction entry.

Using this technique on the table in Figure

[1] Laurence I. Press, "Conversion of Decision Tables to Computer Programs," *Communications of the ACM,* June 1965, 8, No. 6, pp. 385–390.

	R1	R2	R3	R4	EL
C1	Y	N	Y	N	
C2	N	–	Y	N	
C3	Y	Y	–	N	

Figure A3-1a Sample table.

A3-1a, the testing sequence in Figure A3-1b is derived.

To minimize run time for the resultant program, the rules should be ordered on the frequency in which they are expected to be selected (the most common first). However, it would still be necessary to make a large number of tests before isolating most rules since the result of interrogating a condition in one rule is not "remembered" and the same condition may be interrogated many times at a high cost in run time.

This algorithm produces a program with as many tests as there are nonindifferent *entries* in the original table and consequently uses a large amount of core storage.

Despite its failure to minimize either core storage or run time, the algorithm accurately translates the table and might be used solely on the basis of its simplicity and speed of implementation. However, at a small price in simplicity, the rule mask technique proposed by Kirk[2] yields a radically more efficient table translation.

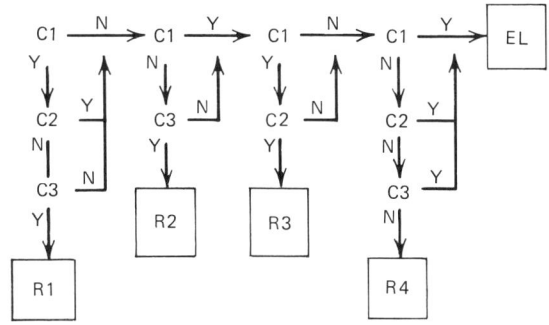

Figure A3-1b Testing sequence of sample table.

[2] H. W. Kirk, "Use of Decision Tables in Computer Programming," *Communications of the ACM,* Jan. 1965, 8, No. 1, pp. 41–43.

162 APPENDIX III

The Simple Rule Mask Technique of Kirk[3]

Kirk describes a rule mask technique, first used in 1958, which consists of the following four steps:

Step 1: Prepare a binary image of the condition matrix of the table by placing a "1" in each position in which the original table has a "Y" and a "0" in all other positions. The resulting matrix is called the "table matrix" and is illustrated in Figure A3-2a for the table shown in Figure A3-1a. This matrix will be scanned until a rule is satisfied.

Step 2: Prepare a binary "data vector" from the actual transaction entry by placing a "1" in each true condition position and a "0" in all other positions. This vector will be used to scan the "table matrix." A sample transaction entry and its "data vector" are shown in Figure A3-2b.

Step 3: Since all indifference entries have been mapped into the "table matrix" as zeros, it is necessary to build a "masking matrix" to screen out nonpertinent entries from the "data vector" before it is used to scan the "table matrix." As illustrated in Figure A3-2c, the "masking matrix" is constructed by placing a "0" in each position in which the original table had an indifference entry and a "1" in all other positions.

[3] *Ibid.*

	R1	R2	R3	R4
C1	1	0	1	0
C2	0	0	1	0
C3	1	1	0	0

Figure A3-2a   Table matrix of sample table.

Condition 1 – False | 0 |
Condition 2 – True | 1 |
Condition 3 – True | 1 |

Figure A3-2b   Theoretical transaction and its corresponding data vector.

	R1	R2	R3	R4
C1	1	1	1	1
C2	1	0	1	1
C3	1	1	0	1

Figure A3-2c   Masking matrix of sample table.

Step 4: The actual scan is made rule by rule. The first rule vector of the masking matrix is logically multiplied by the data vector to eliminate that rule's nonpertinent conditions from the data vector. The result is then compared to the first rule vector of the table matrix. If the two are equal, the rule is satisfied. If not, the scan proceeds to the next rule.

For the table and transaction above, the following illustrates step 4:

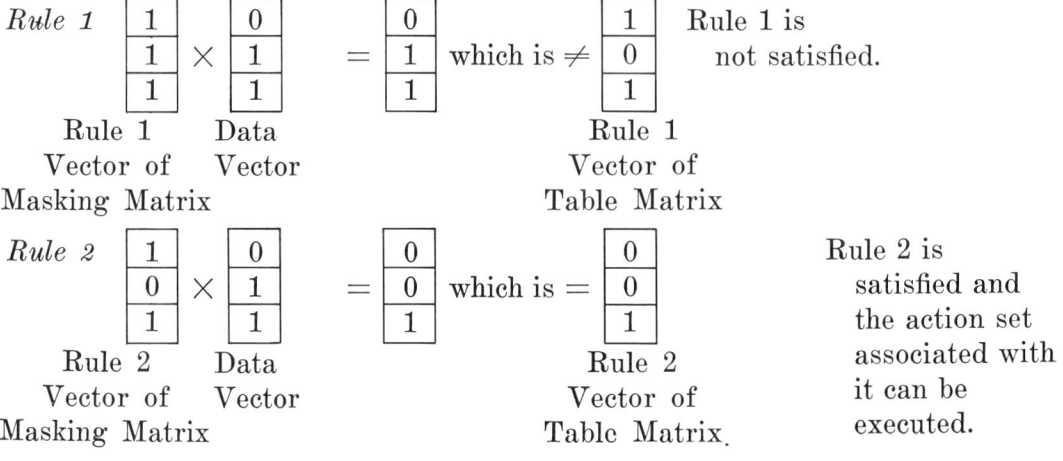

$f_j/\Sigma t_i$		2.7	2.5	2.2	2.3		
Σt_i		13	6	9	13		
f_j		35	15	20	30		
		R1	R2	R3	R4	t_i	Σf_j	$\Sigma f_j/t_i$
	C1	Y	N	Y	N	2	100	50
	C2	N	–	Y	N	7	85	12.1
	C3	Y	Y	–	N	4	80	20

Assume $t = 1$ for multiplying the data vector by the masking vector and comparing the result to the table vector.

Figure A3-3 Sample table of Figure A3-1a, with parameters and calculations for King's Algorithm.

This technique has the advantage of testing each condition only once when setting up the data vector and therefore reduces both core storage and run time over the straight scan technique. The masks themselves are binary and require very little core storage. Run time can be further reduced by arranging the rules in order of the frequency with which they are expected to be selected. It is still a simple algorithm which lends itself to rapid implementation and high-speed translation.

The disadvantages are discussed by Kirk himself: "The technique . . . requires the computer to sense all conditions prior to scanning the decision rules. This could prove to be quite wasteful of computer [run time] in cases where there are a great many conditions and yet the rules most likely to be satisfied have only a few pertinent conditions."[4] He suggests circumventing the problem by splitting the table into two or more pieces and placing those rules with few conditions in separate tables to be interrogated first. P.J.H. King,[5] however, claims that this segmentation would detract from the documentation value of the decision table and that it is unnecessary using his "interrupted rule mask technique."

The Interrupted Rule Mask Technique of King[6]

P.J.H. King departs from Kirk's simple or "uninterrupted" rule-mask technique and shows it can be improved by interruptions.

Using the table in Figure A3-1a and assigning evaluation times (t_i's) to each conditional interrogation and frequencies of occurrence (f_j's) to each rule, as illustrated in Figure A3-3, the computer run time required to process 100 transactions via Kirk's method can be calculated.

Rule 1 (2+7+4+1)   × 35 = 490
Rule 4 (2+7+4+1+1)   × 30 = 450
Rule 3 (2+7+4+1+1+1)   × 20 = 320
Rule 2 (2+7+4+1+1+1+1)   × 15 = 255

Total Run Time 1515

King proposes testing a subset of the conditions and then testing all rules with only those conditions pertinent before doing additional condition testing; for example, C1 and C2 are the only conditions pertinent for R3 and when those two conditions have been evaluated, R3 can be tested immediately. The total run time required to process the same table under this method can be calculated and shows a marked decrease. (Note that if two rules contain the same pertinent conditions, they are tested in descending order of rule frequency.)

---

[4] *Ibid.*, p. 43.
[5] P. J. H. King, "Conversion of Decision Tables to Computer Programs by Rule Mask Techniques," *Communications of the ACM,* Nov. 1966, 9, No. 11, pp. 796–801.

[6] *Ibid.*

164 APPENDIX III

Rule 3 (2+7+1) × 20 = 200
Rule 1 (2+7+1+4+1) × 35 = 525
Rule 4 (2+7+1+4+1+1) × 30 = 480
Rule 2 (2+7+1+4+1+1+1) × 15 = 255
─────
Total Run Time 1460

King has developed four strategies for finding condition and rule testing sequences. The four strategies usually yield different results and the one that produces a testing sequence with the lowest total run time is then selected for use in generating the translated code. The strategies are not guaranteed to produce the optimal testing sequence in all cases but, since enumeration of all possible sequences is not feasible, they are advanced as "sensible ad hoc procedures."[7]

Strategy A tests the *conditions* in descending order of magnitude of relevance frequency (Σf_j). This is based on the supposition that it may be best to evaluate first those conditions most likely to be pertinent. This results in a testing sequence C1-C2-R3-C3-R1-R4-R2 for the table in Figure A3-3 and yields a total run time of *1460* as demonstrated above.

Strategy B tests the *conditions* in descending order of $\Sigma f_j/t_i$. This is based on the supposition that it may be best to evaluate first those conditions with the shortest evaluation times even though they may be less likely to be pertinent. This results in a testing sequence C1-C3-R2-C2-R1-R4-R3 for the table in Figure A3-3 and yields a total run time of *1450*.

Strategy C tests the *rules* in descending order of frequency, evaluating conditions only when they become necessary for testing the rule. This results in a testing sequence C1-C2-C3-R1-R4-R3-R2 for the table in Figure A3-3 and yields a total run time of *1515*.

Strategy D tests the *rules* in descending order of $f_j/\Sigma t_i$. This is based on the supposition that it may be best to test first those rules with relatively shorter condition evaluation times even though they may have lower frequencies. This results in

[7] *Ibid.,* p. 799.

a testing sequence C1-C2-C3-R1-R4-R3-R2 for the table in Figure A3-3 and a total run time of *1515*.

For the table in Figure A3-3, the best testing sequence is derived by using Strategy B which yields a total test time of 1450. This strategy would then be used to translate the decision table into a machine-executable sequence of instructions.

King's technique offers a marked savings in computer run time, especially on large tables, even though it does not necessarily produce an optimal testing procedure for every table. This advantage, unfortunately, has a high price in terms of other efficiency criteria.

The core storage used by the translated program will exceed that of Kirk's translation because of the increased complexity of the branching structure. The algorithm is more complex and consequently would be more difficult and time-consuming to implement. The time required to make the actual table translation would increase accordingly.

King's algorithm relies completely on the user-supplied t_i's and f_j's; these must be accurately estimated or the process deteriorates. Since frequency figures may not remain stable over time, periodic re-evaluation of these figures is advisable and the tables should be retranslated whenever there is a material change.

The Optimal Search Technique of Verhelst[8]

King's algorithm does not *guarantee* an optimal testing sequence requiring the minimum possible total run time since he concludes that enumerating all possible testing sequences is not feasible.

M. Verhelst, in a recent article, proposes an algorithm that does guarantee an optimal testing sequence. This "optimal search technique" *selectively* enumerates *all* possible sequences and their

[8] M. Verhelst, "Procedures for Finding Optimal and Near Optimal Test Sequences for Applying Rule Mask Techniques in Object Programs Derived from Decision Tables," *IAG Quarterly,* Amsterdam, 1968, No. 1, pp. 47–65.

total run time by deleting, at certain stages of the enumeration, those groups of sequences which cannot possibly contain the optimal solution.

This is a rather specialized, complex technique that would notably increase table translation time. It would probably only be considered for use if optimal run time is absolutely essential to the user.

The "2-by-2 Method" of Verhelst[9]

In addition to the "optimal search technique," Verhelst's article also proposes a shorter method called the "2-by-2 method." It is not guaranteed to produce an optimal testing sequence but it does produce a *near* optimal solution at a much lower cost than that of the "optimal search technique."

Briefly, the "2-by-2 method" considers all 2-rule subtables of the original table and determines the optimal rule testing sequence for each of these small tables by the exhaustive search method. These optimal partial sequences are then combined to build a total rule testing sequence for the table by honoring the order of rules in the partial sequences. Finally, the conditions are inserted to produce a total condition and rule testing sequence.

Although it would require less table translation time than the "optimal search technique," Verhelst's "2-by-2 method" would also probably be considered only if optimal or near optimal run time is critical.

CONDITION TESTING AND NETWORK TECHNIQUES

The Two Algorithms of Montalbano[10]

M. Montalbano's paper seems to have provided the initial thinking and impetus for all subsequent work on "network technique" algorithms.

[9] *Ibid.*
[10] M. Montalbano, "Tables, Flow Charts, and Program Logic," *IBM Systems Journal,* Sept. 1962, pp. 51–63.

	R1	R2	R3	R4
C1 =	1	1	2	1
C2 =	3	4	3	3
C3 =	1	1	2	2

Figure A3-4 Sample table.

He proposes two algorithms, the "quick-rule method" and the "delayed-rule method," which will translate a decision table into a machine-executable sequence of test and branch instructions. Uniquely, these algorithms work directly on extended entry decision tables.

The "quick-rule method" is designed to minimize the number of branching instructions in the translated program (i.e., to minimize core storage usage) and "to make as soon as possible those tests which will isolate a rule."[11]

Using the table in Figure A3-4, the quick-rule method is illustrated in Figure A3-5. First, in part (a) of Figure A3-5, the condition portion of the table is listed and a row count matrix is constructed to its right by counting the number of occurrences of each *value* in the condition entries of each row (e.g., in the first row, there are three 1's and one 2). Next, the smallest nonzero number in the row count matrix is found and the conditional interrogations associated with this number are made. In Figure A3-5, part (a), the smallest nonzero number in the row count matrix is "1" which occurs twice. These two interrogations are illustrated in the first two blocks of the flow chart in Figure A3-6 and isolate Rules 3 and 2. These rules are now eliminated and the remaining condition portion of the table is listed, in part (b) of Figure A3-5 along with the row count matrix constructed from it. Condition 3 is found to have the only two nonzero low values in this new row count matrix and can be interrogated for equality to either 1 or 2 to isolate the last two rules. The complete flow chart shown in Figure A3-6 is the final result of the quick-rule method as applied to this table.

[11] *Ibid.,* p. 55.

	Original Table					Row Count Matrix for Original Table				
	R1	R2	R3	R4		1	2	3	4	
	1	1	2	1		3	①	0	0	
	3	4	3	3		0	0	3	①	(a)
	1	1	2	2		2	2	0	0	

Subtable 1		Row Count Matrix for Subtable 1				Subtable 2		Row Count Matrix for Subtable 2			
R2	R3	1	2	3	4	R1	R4	1	2	3	
1	②	1	1	0	0	1	1	2	0	0	
④	3	0	0	1	1	3	3	0	0	2	(b)
1	2	1	1	0	0	1	2	①	①	0	

Figure A3-5   Sample "quick-rule method."

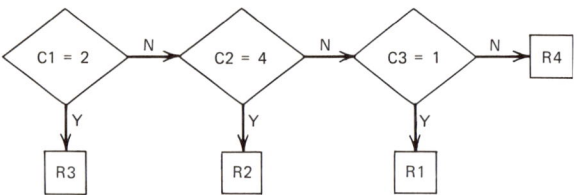

Figure A3-6   Sample "Quick-Rule Method" translation.

The delayed-rule method is designed to minimize the average number of *executed* instructions in the translated program and "to delay as long as possible the tests which isolate rules."[12]

Using the same table, the delayed-rule method is illustrated in Figure A3-7. Here, the row count matrix is searched for a conditional interrogation which will divide the table into *two*

[12] *Idem.*

pieces as *equal* in size as possible and results in the complete flow chart shown in Figure A3-8.

Assuming all rules in this table have an equal frequency, the quick-rule method results in an average execution of 2.5 instructions per transaction while the delayed-rule method has an average of 2.0 instructions executed per transaction. Each requires three conditional test instructions and therefore both use the same amount of core storage.

If rule frequencies were not equal and the relative frequencies were known, the delayed-rule method could be modified to use a row count matrix "weighted" by the rule frequencies. It would then attempt at every step to divide the table into two groups of as nearly equal "weight" as possible.

The table used above had only simple rules, that is, rules in which each entry is explicitly stated. Montalbano considers compound rules, that

	Original Table					Row Count Matrix for Original Table				
	R1	R2	R3	R4		1	2	3	4	
	1	1	2	1		3	1	0	0	
	3	4	3	3		0	0	3	1	(a)
	1	1	2	2		②	②	0	0	

Subtable 1		Row Count Matrix for Subtable 1				Subtable 2		Row Count Matrix for Subtable 2				
R1	R2	1	2	3	4	R3	R4	1	2	3	4	
1	1	2	0	0	0	2	1	①	①	0	0	
3	4	0	0	①	①	3	3	0	0	2	0	(b)
1	1	2	0	0	0	2	2	0	2	0	0	

Figure A3-7   Sample "delayed-rule method."

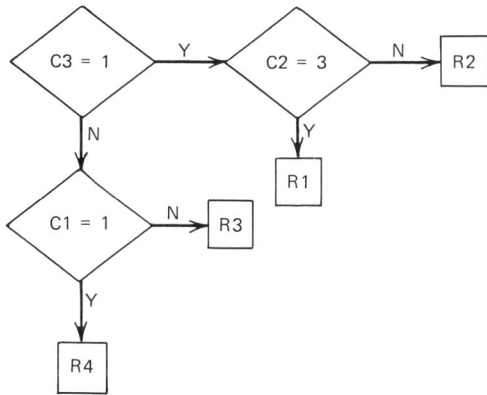

Figure A3-8 Sample "Delayed-Rule Method" translation.

is, rules in which "don't-care" entries appear, and the Else-rule in a "modified delayed-rule method."

He notes that "the effect of a "don't-care" indicator is to consolidate several [simple rules] into one [column]—the number of [simple rules] depending on the number of alternatives possible for the logical variable to which the 'don't-care' applies."[13] If there are two or more "don't care" entries in the same rule, the number of simple rules implied is equal to the *product* of the number of alternatives possible for *each* logical variable in which there is a "don't care." This count of simple rules is called the "column weight" and is indicated for the table in Figure A3-9 at the foot of each column. Since each table has S simple rules (where $S = p_1 \times p_2 \times p_3 \times \ldots \times p_n$, p_i is the number of logical alternatives for condition i, and n is the number of conditions in the table)

[13] *Ibid.*, p. 58.

the Else rule must have a column weight equal to S less the sum of the column weights of all other rules.

Using this information, the "modified delayed-rule method" is developed. It calculates the row count matrix (or weighted row count matrix) for all rules except the Else rule by *weighing* the number of occurrences of each *value* in the condition entries of each row by the column weight of the column in which it appears. In the first row, for example, there are three 2's with column weights of 1, 1 and 2 so that the entry in the row count matrix for row 1, value 2 is 4; also in the first row, there is one dash with a column weight of 8 so that the row 1 dash entry value in the row count matrix is 8. The initial row count matrix for the table in Figure A3-9 is shown in Figure A3-10.

The "delayed-rule method" can now be applied as before except, in addition to finding a conditional interrogation which divides the table into two parts as equal in size as possible, any row with a "don't care" entry must be avoided as long as possible. In Figure A3-10, row two has the only zero in the dash column so only tests in that row will be considered. Testing condition 2 for either a 2 or a 3 will come closest to splitting the table into two equal parts and either test can be made.

Montalbano's methods serve as the basis for Pollack's[14] algorithms which are described below.

[14] Solomon L. Pollack, "Conversion of Limited-Entry Decision Tables to Computer Programs," *Communications of the ACM,* Nov. 1965, **8**, No. 11, pp. 677–682.

	R1	R2	R3	R4	R5	R6	R7	EL	
C1 =	1	1	1	2	2	2	–		2 possible alternatives
C2 =	1	1	2	2	2	2	3		3 possible alternatives
C3 =	–	–	–	1	1	2	–		2 possible alternatives
C4 =	1	2	–	1	2	–	–		2 possible alternatives
Column weight	(2)	(2)	(4)	(1)	(1)	(2)	(8)	(4)	($S = 2 \times 3 \times 2 \times 2 = 24$)

Figure A3-9 Table to illustrate calculation of column weight.

-	1	2	3
8	8	4	0
0	4	8	8
16	2	2	0
14	3	3	0

Minimum dash row

Figure A3-10 Row count matrix for table in Figure A3-9.

The Procedure of Egler[15]

This algorithm handles limited entry tables only and attempts to *simultaneously* minimize translation time, core storage used by the translated program and run time required to execute the translated program. It is based on the supposition that those conditions used most often (i.e., those with the fewest "don't-care" entries) should be tested first.

M. Montalbano[16] demonstrates that the pro-

---

[15] J. F. Egler, "A Procedure for Converting Logic Table Conditions into an Efficient Sequence of Test Instructions," *Communications of the ACM,* Sept. 1963, **6**, No. 9, pp. 510–514.

[16] M. Montalbano, "Egler's Procedure Refuted," *Communications of the ACM,* Jan. 1964, **7**, No. 1, p. 1.

	R1	R2	R3	R4	R5	R6	R7	R8	R9
C1	Y	N	-	-	Y	Y	N	-	N
C2	-	N	N	Y	N	Y	Y	Y	Y
C3	Y	Y	N	N	Y	Y	Y	N	Y
C4	N	-	N	N	Y	Y	N	Y	Y

Figure A3-11 Table to illustrate algorithm failure.

cedure "will not, in general, do what is claimed" by using the table in Figure A3-11.

According to Egler's procedure (as illustrated in Figure A3-12):

1. All N's in the original table's condition matrix are replaced by 0's; all Y's by 1's and all "don't-cares" by blanks.

2. The number of 0's *and* 1's in each row is calculated and called the "Y-N-Count."

3. The original table's condition rows are sorted into descending order of "Y-N-Count."

4. The rule columns are then sorted into numerical order, row by row, treating blanks as lower than zeros.

5. The conditions are tested in order from top to bottom of the reformatted table; that is, C3-C2-C4-C1.

	R1	R2	R3	R4	R5	R6	R7	R8	R9	Y-N-Count
C1	1	0			1	1	0		0	6
C2		0	0	1	0	1	1	1	1	8
C3	1	1	0	0	1	1	1	0	1	9
C4	0		0	0	1	1	0	1	1	7

Original Table – Steps 1 & 2

	R3	R4	R8	R1	R2	R5	R7	R9	R6	Y-N-Count
C3	0	0	0	1	1	1	1	1	1	9
C2	0	1	1		0	0	1	1	1	8
C4		0	1	0		1	0	1	1	7
C1				1	0	1	0	0	1	6

Reformatted Table – Steps 3 & 4

Figure A3-12 Conversion of table in Figure A3-11.

Montalbano notes that after testing C3, the *true* subtable is no longer in descending order of condition usage (Y-N-Count). Testing C1 at this point would produce better results but Egler's procedure selects C2.

Other failures of Egler's procedure are its inability to handle rule frequencies and the Else-rule. These failures are avoided by Pollack's two algorithms.

The Two Algorithms of Pollack[17]

Pollack's algorithms are based on the previously described work of Montalbano.[18] They convert limited-entry decision tables to a machine-executable sequence of instructions, automatically handle the Else-rule and isolate any redundant or contradictory decision rules during the conversion process. The following two algorithms have been updated to handle dependent conditions and improve efficiency. The improved algorithm is illustrated in the text. For illustrative purposes in this appendix, the original algorithms are used.

Algorithm 1 minimizes the number of comparison instructions in the translated program thus minimizing its core storage usage. Algorithm 2 minimizes the number of comparison instructions *executed* in the translated program thus minimizing its computer run time. Any rule not explicitly stated or implied in the table is assumed to belong to an Else-rule which need not be stated. The two algorithms *require* unambiguous tables.

Before beginning the actual translation, the number of rules in the original table that result in the same sequence of actions must be minimized. Any two rules with the same action set which differ in condition entries by only one Y-N pair may be combined into a single rule with a "don't-care" entry in the Y-N pair row.

Generally, Pollack's conversion of a table to a sequence of test instructions follows the following steps:

1. A single condition row of the original table is selected (the selection criteria are specified in the two algorithms). This condition becomes the first comparison of the testing sequence.

2. The original table is split on the selected condition into two subtables or into a subtable and a rule or into two rules. Each subtable is associated with one branch of the selected conditional interrogation and each is one row smaller than the original table since it lacks the selected condition row.

3. From *each* subtable, a single condition row is selected and becomes the next comparison of the testing sequence.

4. The process is continued until all rules have been isolated or a contradiction or redundancy is found.

ALGORITHM 1. "The objective is to convert a decision table to a computer program and have this program use the minimum number of storage locations."[19] The procedure is illustrated in Figure A3-13, the resulting test sequence in Figure A3-14.

Step 1: Check the table for redundancies and contradictions. If two rules do not contain at least one row where one rule has a Y entry and the other has an N entry, the two rules are either redundant or contradictory. They are redundant if they have the same action set, contradictory if they do not.

Pollack suggests doing the redundancy/contradiction test only when a table has been reduced to one row. At this point, the table is unambiguous *only* if there is only one rule or there are two rules, one with a Y entry the other with an N.

Step 2: Calculate the "column count" (cc) for each *rule*. It is equal to 2^r where r is the number of dashes ("don't care" entries) in the rule.

Step 3: Calculate the "dash count" (dc) for each *row*. It is equal to the sum of the column counts of all rules that have a dash entry in the row.

Step 4: Select the condition to be interrogated (denoted C_k). If there is a single row with a *minimum* dash count, this row is C_k. If two or

---

[17] Pollack, *loc. cit.*
[18] M. Montalbano, "Tables," *loc. cit.*
[19] Pollack, *op. cit.,* p. 679.

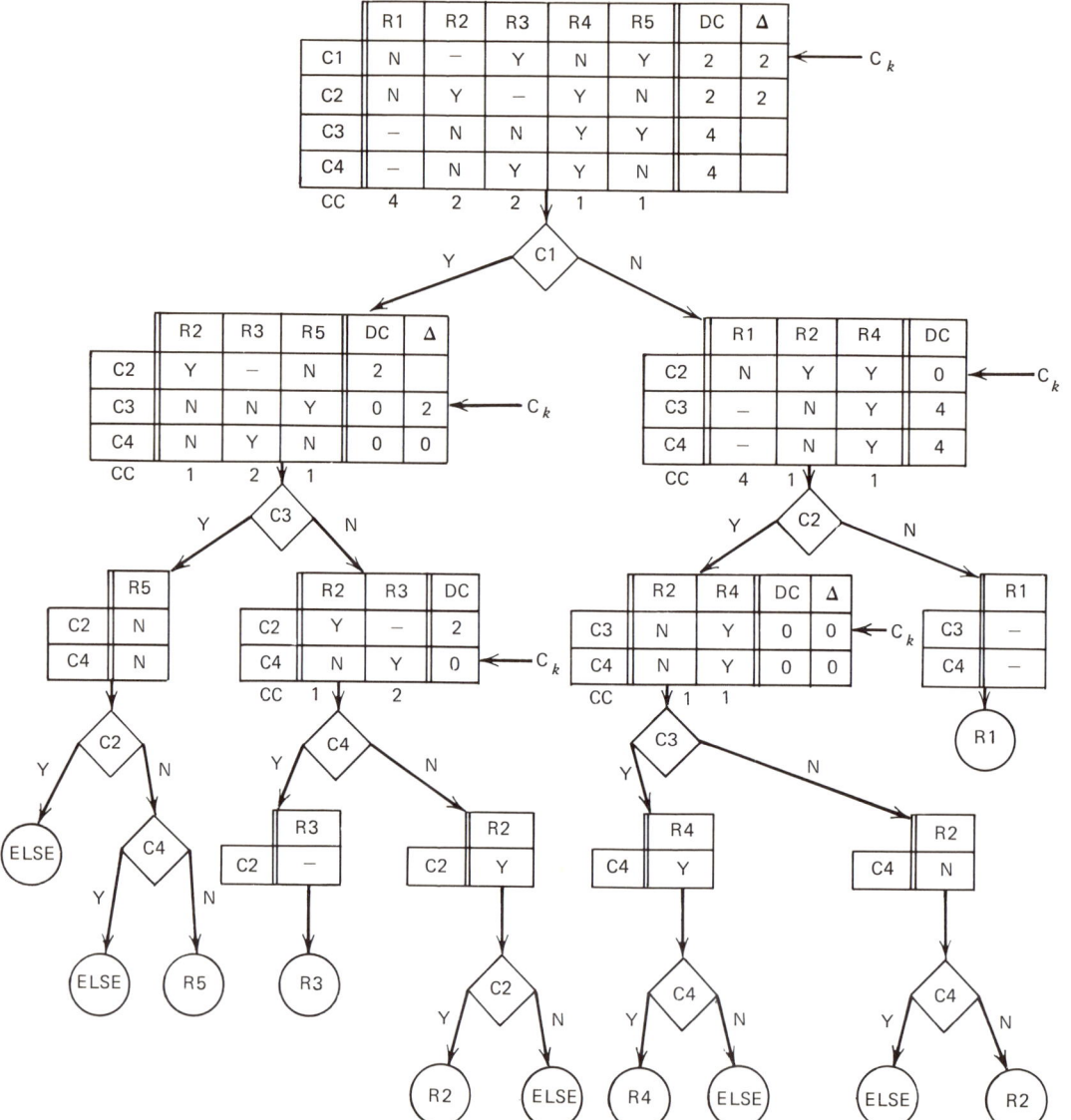

Figure A3-13 Translation of a decision table using Algorithm 1.

more rows tie for minimum dash count, select from among them the row with *maximum* "delta" (Δ). Delta, for each row, is equal to the absolute value of the difference between the "Y-count" and the "N-count" for that row. The Y-count (or N-count) is the sum of the column counts of all rules that have a Y (or N) in the row.

Step 5: Discriminate on C_k. The two branches of the discrimination each contain a subtable with one or more rules and one row less than the previous table (row k is dropped).

The Y-branch subtable contains all rules from the previous table that had a Y or a dash entry in row k. The N-branch subtable contains all rules that had an N or a dash entry in row k.

Step 6: If the subtable contains more than one rule return to Step 1.

Step 7a: If the subtable has exactly one rule that contains only dashes, that rule has been isolated.

Step 7b: If the subtable has exactly one rule but more than one condition and does not contain

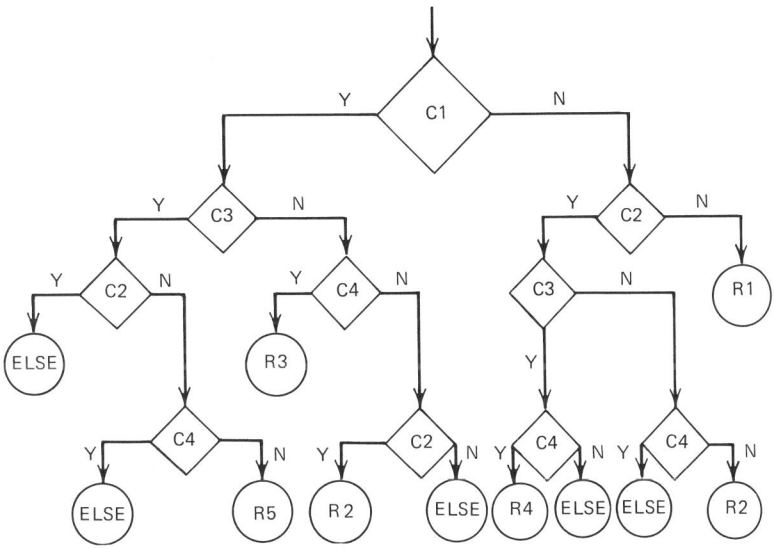

Figure A3-14 Test sequence generated in Figure A3-13.

only dashes, choose any non-dash row and discriminate on it. This will yield a subtable from the satisfied condition and an Else-rule isolation from the opposing branch.

Step 7c: If no subtable is produced, an Else-rule isolation is indicated.

Step 7d: If the subtable has exactly one rule that has one condition with a Y or N entry, discriminate on the condition. The satisfied branch isolates the rule while the opposing branch isolates the Else-rule.

ALGORITHM 2. "The objective is to convert a decision table to a computer program whose comparisons can be executed in minimum time."[20] It requires user-supplied rule frequency figures (f's) that estimate how often each rule will be selected when testing an average sampling of transaction entries. The algorithm assumes that relatively few transactions will satisfy the Else-rule.

The column count, dash count, Y-count, N-count and delta figures are calculated according to the specifications in Algorithm 1. Using the same table in the illustration for Algorithm 1 (in Figure A3-13), the procedure for Algorithm 2 is illustrated in Figure A3-15 and its resulting test sequence in Figure A3-16.

Steps 1 and 2: Same as Algorithm 1.

Step 3: Calculate the "weighted dash count" (WDC) for each *row*. It is equal to the sum of the products of rule frequencies and column counts of all rules that have a dash entry in the row.

Step 4: Select the condition to be interrogated (C_k). If there is a single row with a *minimum* weighted dash count, this row is C_k. If two or more rows tie for a minimum weighted dash count, select from among them the row with *minimum* delta. If there is still a tie, select from among the minimum delta rows the row with the *minimum* dash count. (The dash count test can save core without affecting run time.)

Steps 5, 6, and 7: Same as Algorithm 1.

Each algorithm produces a testing sequence with ten conditional interrogations. The sequence produced by Algorithm 2, however, has a lower total test time for a set of transactions that satisfy rules in accordance with the supplied frequencies.

Assuming each conditional interrogation takes one time unit, the total test time required by each rule is equal to the product of the number of conditional interrogations required to isolate the rule and the rule's frequency. For each translation the

[20] *Ibid.*, p. 681.

172 APPENDIX III

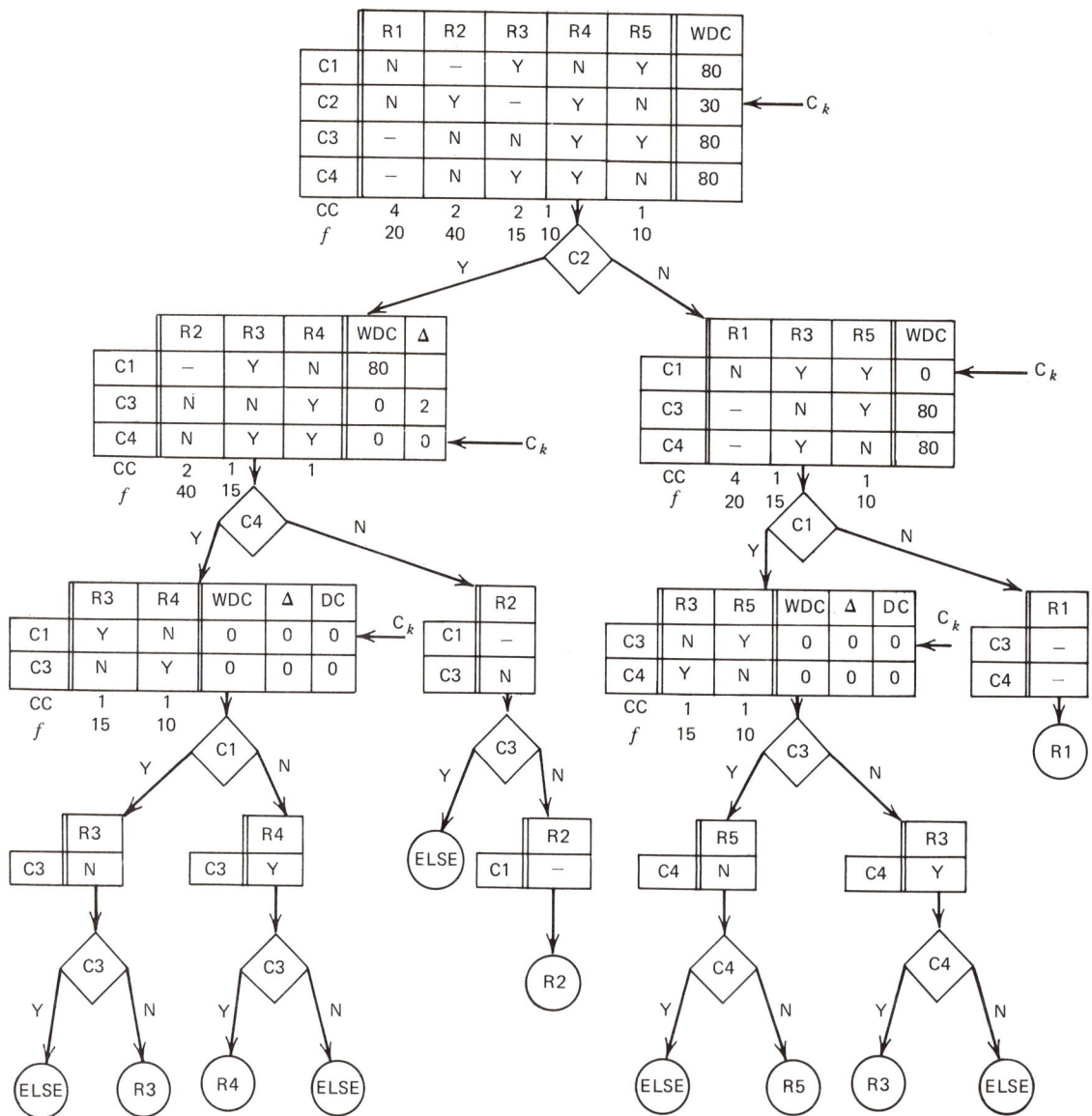

Figure A3-15   Translation of a decision table using Algorithm 2.

total time required to execute one hundred transactions is calculated below:

	Algorithm 1	Algorithm 2
Rule 1	$2 \times 20 = 40$	$2 \times 20 = 40$
Rule 2	$4 \times 40 = 160$	$3 \times 40 = 120$
Rule 3	$3 \times 15 = 45$	$4 \times 15 = 60$
Rule 4	$4 \times 10 = 40$	$4 \times 10 = 40$
Rule 5	$4 \times 10 = 40$	$4 \times 10 = 40$
Else	$3.8 \times 5 = 19$	$3.8 \times 5 = 19$
Total test time	$= 344$	$= 319$

Pollack offers no proof of the optimality of his algorithms and expresses the hope that "others will develop the necessary proofs or offer counterexamples to prove that the algorithms fail."[21]

Sprague's Counterexample, Comments and Criticism[22]

Commenting on Pollack's article,[23] Sprague demonstrates that "neither algorithm always ac-

[21] *Ibid.*, p. 679.
[22] V. G. Sprague, "On Storage Space of Decision Tables," *Communications of the ACM,* May 1966, 9, No. 5, p. 319.
[23] Pollack, *loc. cit.*

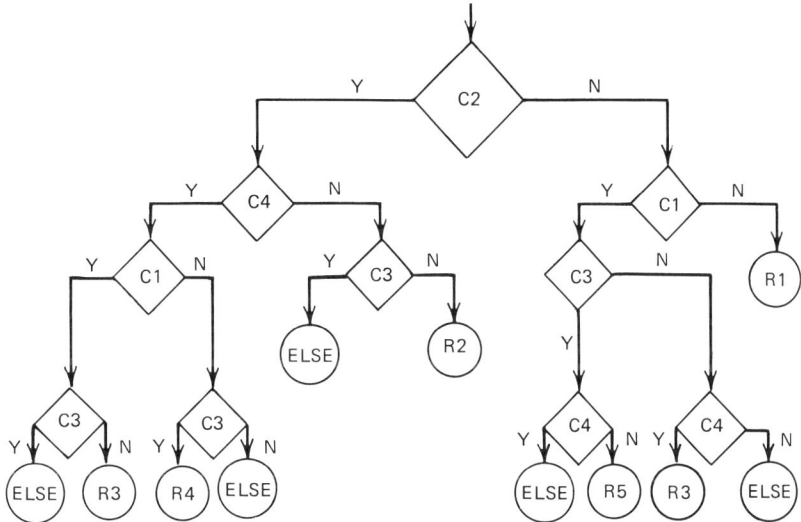

Figure A3-16   Test sequence generated in Figure A3-15.

	R1	R2	R3	R4	R5
C1	N	–	Y	N	Y
C2	N	Y	–	Y	N
C3	–	N	N	Y	Y
C4	–	N	Y	Y	N

Figure A3-17   Sample decision table to illustrate failure of algorithm 1.

complishes the desired minimization."[24] Using the table in Figure A3-17, he notes that Pollack's Algorithm 1 yields ten branches. If, however, an initial discrimination is made on C4 and then Algorithm 1 is applied, the resulting test sequence contains only nine branches.

Using Karnaugh graphs, Sprague develops a geometric interpretation of Algorithm 1 and states that "the reduction of a table to tree-form now becomes equivalent to a corresponding problem in multidimensional geometry."[25] Using this geometric interpretation, he itemizes his criticisms of Algorithm 1:

1. "Action 'areas' are the important parameters and the rules . . . are just a clumsy method for specifying the action areas."[26] He feels that a true optimizing algorithm *cannot* concentrate exclusively on the rules and ignore the actions.

2. The dash count and delta are statistical and ignore the geometric relationships of the action areas.

3. The Else rule must be treated as any other action area since it too occupies a geometric area.

From the geometric approach, Sprague concludes that "Algorithm 1 becomes increasingly unreliable with increasing number of conditions."[27]

Switching to a Boolean interpretation of the core minimization problem, he notices another possibility for minimization. "If any subtable of an expansion-tree is identical to or can be transformed by Boolean operations into any other subtable in the tree, then these two subtables may be identified as a single subtable."[28] Pollack[29] agrees that checking for equal pairs of subtables would be a valuable addition to his algorithms if the user is willing to allow the additional cost during the table translation.

---

[24] Sprague, *loc. cit.*
[25] *Idem.*
[26] *Idem.*
[27] *Idem.*
[28] *Idem.*
[29] Solomon L. Pollack, "Letter to the Editor," *Communications of the ACM*, May 1966, **9**, No. 5, p. 320.

The Algorithms of Reinwald and Soland[30]

The most rigorous work on translating decision tables to an optimal branching sequence has been done by Reinwald and Soland. They have developed two very general algorithms that *guarantee* optimality of the resulting test sequence. One algorithm minimizes run time and the second minimizes core storage. Furthermore, the authors claim that the two algorithms can be combined to yield a testing sequence that minimizes a total cost function defined on both core usage and run time.

The algorithms are quite complex and suitable only for machine implementation. Their implementation is discussed in the articles.

Because of limited space, the algorithms are not described here. The interested reader is referred to the cited papers.

CONCLUSION

Both "network technique" and "rule mask" algorithms have been developed to a high level of sophistication. The sophisticated network algorithms of Reinwald and Soland and the sophisticated rule mask algorithms of Verhelst yield optimal results but are complex, unsuitable for manual processing, difficult to implement in a preprocessor, and notably increase table translation time. The simpler algorithms "should not therefore be dismissed out of hand because they do not always achieve the absolute best. It may be that the relative simplicity of these procedures compensates for their not always producing the optimum, for even when they produce a nonoptimum solution it is likely to be a fairly good one."[31]

---

[30] L. T. Reinwald and R. M. Soland, "Conversion of Limited-Entry Decision Tables to Optimal Computer Programs I: Minimum Average Processing Time," *Journal of the Association for Computing Machinery,* July 1966, **13**, No. 3, pp. 339–358.

———, "Conversion of Limited-Entry Decision Tables to Optimal Computer Programs II: Minimum Storage Requirement," *Journal of the Association for Computing Machinery,* Oct. 1967, **14**, No. 4, pp. 742–758.

[31] P. J. H. King, "Decision Tables," *The Computer Journal,* Aug. 1967, **10**, No. 2, pp. 135–142.

Index

A, symbol for actions, 19
  action identifier, 88
Abbreviations, 66
ACM, 5
  Los Angeles Chapter of, 5
Actions, 3, 7, 13, 39–44, 78
  area, 8
  checklist, 94
  combination of like, 44
  enter table from, 44
  execution of, from conditions, 37, 44
  extended entry, 9, 41
  grouping of, 50, 113
  initialization as, 49
  limited entry, 40, 41
  marker X, 8, 40
  optimization of, 50
  redundancy and contradiciton, 54
  row, 86
    check of, 99, 110
    conditional statements in, 42, 43
    operational statements in, 40
    replication or repetition of, 42, 43
  rule entry, 41
  section, 40
  sequence, 39, 40
    different for various rules, 42
    normal, 42; *see also* Action Sequence Numbers
    special features, 44
  stub, 40, 41
  subroutine, 44
  see also Operational statements
Action Sequence Numbers, 42
  extended entry, 42
  indicators, 42
  limited entry, 42
  mixed usage, 42
ADC, 132
Adding tables, 111
Additions to a table, 78
AF: *see* AND–Functions

ALGOL, 63, 66
Algorithms, conversion, 5; *see also* Decomposition
  definition of, 132
  minimization of number of decision branches (Core optimization), 134
  presented as a decision table, 135
American National Standards Institute (ANSI), 65, 66, 154
Analysis of Systems, 13
AND, Boolean, 16
  COBOL, 158
  connected conditions, 27, 28
  logical operator, 3, 8
AND-Functions, 16–19, 22–24
  dependent, 17, 22–24
  independent, 17, 22, 24
  mixed, 21, 23, 146
  pure, 19, 21, 23, 146
  theorems, proofs, 195
ANS, 65, 66, 154
APAR, 113
Applications used as examples, age and sex, 32
  bond-deduction register, 88
  editing, 18
  FICA, 35, 115
  inventory, 33
  manufacturing, 64
  modulo numbers, 41
  order filling, 43
  payroll, 74, 79
  quality control, 31
  routings, 118
  spaces detection, 36
  verbs, 30
  work assignment, 119
Arithmetic statements, COBOL, 158
ASN: *see* Action Sequence Numbers
Association for Computing Machinery, 5
Assumptions of decision table theory, 13
Asterisk, 16, 18, 21, 30, 34, 46, 50, 55

dollar sign, reversal of assignment, 36, 131
Asterisk/Dollar Count, 132
Asterisk/Dollar rows, summary of checks, 136
Asterisk entry, horizontal effect of, 31, 130, 131
Average logical rule selection path, 45
Axion, underlying, 21
Axioms of decision table theory, 13

B: *see* AND-Functions
BASIC, 66
Basic assumptions of decision tables, 53
Between, as a condition, 31
Bibliography, 139–143
Binary Matrix tables, 55, 79
  extended entries, 56
Binary Yes-No, 28, 29
Boeing Company, 5
Books on decision tables, 6; *see also* Bibliography; Manuals
Boolean, AND, 16
  NOT, 16
  Exclusive OR, 19, 23
  Inclusive OR, 16, 18
  OR, 16, 18
Bounded conditions, don't-cares in, 33
  extended entry conversion to limited entry, 129
  one row, 33
  simulating with traditional entries only, 34
  two rows, 32
  varying limits, 36, 37
Bounded Ranges: *see* Range
Bound Operators, 32, 56, 122
  GT, 32
  GE, 32
  LT, 32
  LE, 32
  NE, 32, 33, 61

INDEX

C, symbol for condition, 14, 17, 88
  states of, 16
  see also Condition row
Canonical form, 23
Capacity of a table, 115, 116
Case study, 119–125
Cathode Ray Tube (CRT) devices, 68
Changes to a table, 78
Checklist, miscellaneous, 94
Chess, similarities to, 129
COBOL, 4, 5, 32, 34, 39, 40, 64, 65, 86, 98, 154
  ANS, 66
  COBOL–61, 5
  COBOL–F, IBM, 66
  IBM S/370, 106
  Introduction to, 154–159
  Macros, 66
CODASYL, 4, 5
  COBOL Journal of Development, 154
Colon Operator: see Substitution operator
Combination, of tables, 79
Commercial programming, 64
Commercial system, 73
Common Business Oriented Language: see COBOL
Communications, inter rule, 44
  inter table, 44
  intra table, 44
Communication-Oriented Decision Table, 4
Communication value of decision table, see Documentation value
Compiler development, 73
Compilers, 98; see also individual names
Completeness, 21, 110
Completeness checking, 58, 75, 123
  bounded entry, 58
  extended entry, 58
  non-overlapping ranges, 59
Complex AND-Function, 19
Complex decision rule, 21
Complex logic, 13
Compound decision rule, 21
Computer programs, using decision tables in, 53, 85–91
Computers, CDC, 1604, 3400, 3600, 5
  GE 225, 3, 5
  IBM 1401, 7040, 7044, 7080, 7090, 7094, 360, 5
  IBM 702, 305, 650, 704, 3
Conceptual correctness, 104
Conditional statements, 39
Condition-name, COBOL, 77
Conditions, 3, 7, 27–37, 78
  actions, executed during evaluation, 37
  addition of, 108, 110
  bounded: see Bounded conditions
  changes to, 110

checklist, 94
completeness of, 75
conjunction of, 27
deferring of, 42
dependence between, 14, 17, 19, 24, 110
dependency, mutual exclusion, 14, 28–30, 32, 48, 53, 54
extended entry, 9, 55
grouping of, 80
implicit combining, 56
independence between, 15, 29, 110
limited entry, 28
nonexclusion, 54
nonrelatedness of, 37, 50
order of, 50
overall analysis of, 6
overlapping dependencies, 14–15
related, 29
  recognition of, 29
removal of, 110
requirements, 15
row, 86
  check of, 99
state of, 29
structure of, 14
stub, expression in, 28
traditional, 29
true or false, 13, 21
Condition variable, 17, 21
  COBOL, 157
Conference on Data Systems Languages, 4, 5
Confidence, level of, 106
Console debugging, 68
Contradiciton, 21, 53
  in completeness checking, 59
  manual check for, 100
  see also Redundancy and Contradiction
Contradictory rules, procedure for handling, 103
Control paths, testing of, 105
Control record, 86
Conversion methods: see Algorithms, conversion
Core, 45
Correctness, verification of, 97

D, symbol for Decision Rule, 19
Dash: see Don't-care
Dash Count, 132
Data, orderly presentation of, 3
Data demand relationships, 83
Data description, COBOL, 155
Data-item to data-item, relation of, 35
Data requirements, 79
DC, symbol, 132
Debugging, of programs, 97–106
Debugging runs, 104
Decision Boxes, 48–50
Decision Logic Translator, 5

Decision Rule, simple, 19
  complex, 24
  see also Rule
Decision Structure Tables, 3
Decision Tables, definition of, 3, 19
  experimental, 4, 5
  falling out of, 41, 91
DECISUS, 86
DECITAB, 44, 85, 86, 111
  and COBOL, 85
Decomposition, 28
Decomposition algorithms, 129–137
  core and run optimization similarities, 46, 48
  core optimization, 46, 47
  decomposition process, 46
  run time optimization, 46, 47
  top-to-bottom decomposition, 47, 48
  see also Algorithms; Translation algorithms
Delayed Rule, technique, 166
Deleting tables, 111
Delta Count, 132
DeMorgan's Theorem, 18
Dependence and independence differences in meaning between conditions and rules, 54, 100
Dependency, condition: see Conditions, dependency
  rule: see Rule dependency
Dependency diagrams, 15
Design of systems, 13
  design phase, as opposed to analysis phase, 73
  steps of, 77
DETAB–X, 4, 5
DETAB–X Seminar, 5
DETAB/65, 5, 6, 86
DETAP, 86
DETOC, 86
DETRAN, 86
Diagnostic run, 104
Dialects and jargons, 76
DISPLAY, COBOL, 106
Ditto, horizontal, 37
Division, of tables, 79
Divisions, COBOL, 154
DO, FORTRAN, 66
Document, Decision Tables as a, 41, 78, 115
Documentation of maintenance changes, 113
Documentation value, of Decision Tables, 4, 61, 75, 109, 116
Dollar Sign entry, 16, 18, 21, 30, 34, 46, 50, 55
Dollar Signs, steps in using to express ranges, 34
Don't-cares, 8, 27, 29, 33, 48, 55, 83
  simple rules, 55
DT, symbol for decision table, 19

E, symbol for inclusive OR-Function, 18

INDEX

Efficiency, aids to, 29
　in rule combination, 61
　in terms of resources, 45
Egler, J. F., 168
Egler, The Procedure of, technique, 168
Ellipsis, COBOL, 63
ELSE, COBOL, 64
　local, 36
　rule, 19, 28, 40, 53, 75
　　and completeness, 58, 59, 109, 110
Else rule, as an error trap, 60
　extraction from, 109
　residual use, 60
　when it is needed, 60
English language, 63, 64
　in system analysis, 64
Entering tables, 44
Entry, 7
　condition, 9
　negation of, 58
　sense of, 56
EQUAL, 36
Equal case, between ranges, 62
Errors, automatic correction of, 104
　detection of and recovery from, 79, 106
　sources of, 94
　undetected, 106, 107
Evaluation of table, 14; see also Rule selection
Evans, Orren Y., 4
Exhaustive testing, 105
EXHIBIT, COBOL, 106
Explicit entries, in rule combination, 56
Explicit requirements, 16, 30
Extended entry, 55
　conversion to limited entry, 28, 29, 36, 56, 130
　endings, 29

Fall through, 41, 91
Features, checklist for selection of, 94
Figurative constants, COBOL, 156
Flowcharts, 3, 6, 46, 47, 75, 78, 85
　macro, 117
Format, check of, 99
　of input, 69
Formula, Formuli, 3
FORTRAB, 5
FORTRAN, 5, 63–65, 86
Frequencies, 86
　changes to distribution of, 111
　frequency, 46, 61
　frequency row, check of, 99
Full table, check of, 100

GE, 3–5
General Electric, 3–5
GENERATE, COBOL, 66
GO TO, COBOL, 65
　FORTRAN, 66
　DEPENDING ON, COBOL, 39

Grammar, 64
Greater Than, as condition, 31
　overlapping, 34

Hand conversion of tables, 77, 85
Historical usage, 29
History, 3–6
Hits: see Rule, hits
Horizontal effects, 130, 131
Host language, 40
Host program, 97, 124, 125
　compilation, 104
　manual check of, 104
　relation of tables to, 98
Hunt Foods and Industries, 4

I, negation of, ϕ, 18, 21, 23, 24
　symbol for don't-care, 16, 21, 24, 28, 30
　see also Don't-cares; Initialization
IBM Corporation, 5
Identical entries, 32
IF, COBOL, 43, 45, 46, 48, 64, 157, 159
　FORTRAN, 65
　logical, 8
　programming, 27, 28
Imperative Procedure, statements, 39
Implicit decision rules, 19
Implicit entries, requirements, 16, 30
　in rule combination, 56
　simple rules, 60
Inclusive OR-Functions, conversion to AND-Functions, 149
Incorrect Tables, examples of, 30, 32, 37
Inefficiency, 28
Inequalities, as conditions, 31
　between data items, correct and incorrect handling, 37
　between multiple data items, 36
　between two data-items, 36
Initialization, bypassing, 44
　changes in, 111
　reexecution of, 44
　row, check of, 99
　statements, 86
Input/Output description, 80
Input/Output statements, COBOL, 156
Input stream, 86
Instructional use of tables, 13, 53
Insurance Company of North America, 5
Interactions between tables, 115–118
Interactive Programming, 68
Interfaces, 78
　information, 82
Interpretive Decision Tables, 86
Interrupted Rule Mask, 163
Iteration, 116

Joint Users Group (JUG), 5

Karnaugh Map, 4

King, P. J. H., 163, 174
Kirk, H. W., 161

Languages, 97
　artificial, 64
　assembly, 66, 86
　extensions and restrictions, 63
　higher level, 66
　machine, 86
　natural, 63, 64
　programming, 64
　scientific and engineering, 65
　standardization of, 66
　used in decision tables, 63–69, 76
　ways of using decision tables with, 86
　see also languages by name
Less than, as conditions, 31
　overlapping, 34
Limited entry, 8, 56
　traditional technique for writing, 29
Linkage between tables, 77
LOBOC, 5
Logical Integrity, 32, 68
　checking, 123
Logical Structures, flowcharting of, 46
Logical Tree, Structure, 46
LOOP, 111, 117
Looping, 88, 116

M: see Mixed AND-Function
Macros, expansion of parametric and nonparametric, 66, 67
Maintenance, 78, 107
　miscellaneous changes, 111
　programmer, 107
　report form, sample, 112
Management Rules, 4, 75
Man-to-man communication, 4, 73, 83
Manual checking, 99
Manuals, handbooks, programmer's guides, 63, 93
Manufacturing processes, 3
　decisions, 4
Mathematics, as a natural language, 64
Mixed AND-Function, 19, 23, 146
Mixed Entry table, 9
Modifications of output, manual, 50
Modules, check of, 105
Monitoring and control of tables, 117
Montalbano, M., 165, 168, 169
MOVE, COBOL, 63, 154, 158
Multi-programming, 68
Mutual exclusion, in conditions: see Conditions, dependency

n, number of conditions, 14, 55
N: see NO
Narrative, 3, 6, 75, 78
　translation to decision table, 121
Network technique, 165

178 INDEX

NO, 8, 16, 18, 21, 28–30, 46, 50, 55
Non-bounded Extended Entry, 32
Non-COBOL operators, 74
Non-overlapping bounded extended entry, conversion to limited entry, 131
NOT=: see NOT EQUAL, NE
Notation, of decision tables, 19
NOT, 16
NOT EQUAL, as condition, 31, 32, 36
Null Requirement: see "φ"

OF, 22
Operational statements, 39–41
   period at end, 40
   sequence of, 39–41
Operational steps: see Operational statements
Optimal Search, technique, 164
Optimization, 45–51
   actions, 113
   and decomposition, 45
   basis of, 28
   core, 45, 49
     flowchart, 46
   manual, 50
   run time, 45
   run time flowchart, 46
   special features, 44
   types of, 45
OR, Boolean: see Boolean Inclusive OR; Boolean Exclusive OR
OR, logical operator, 3
   relationship between conditions, 27
OR-Functions, 19, 22–24
   dependent, 22
   independent, 22
OS Operating System, IBM, 66
Overlapping dependencies, in conditions, 14
Overlapping ranges, rule uniqueness, 36
   using traditional conditions only, 33

P: see Pure AND-Function
Parallel runs, 106
PERFORM, COBOL, 44, 65, 77, 156
Performing Tables, 44
PL/I, 66, 86
Pollack, Solomon L., 167, 169, 173
   the two algorithms of, technique, 169
Postprocessing, 50
Preprocessors, combined rules, 68
   frequencies, 49
   possible improvements in, 50
   properties of, 6, 29, 32, 33, 40, 44, 45, 61, 85
Press, Laurence I., 161
Problem analysis form, 112
Problems, ways of looking at, 6
PROCEDURE DIVISION, COBOL, 98

Programming, 77
Programming Courses, 6
Proprietary Decision Table Processors, 5
Pure AND-Function, 146

Quick-rule, technique, 165

RAND Corporation, 5
Ranges, extremes of, 132
   full overlapping, 35
   manual check for, 101
   non-contignous, 62
   non-overlapping, 56
   of values, as condition, 31
   rc, rc': see Rule count
Readability of a table, 79
Redundancy, 21
   in completeness checking, 59
   manual check for, 100
   Redundant comparisons, 28, 30
   Redundant rules, elimination of, 103
   procedure for handling, 102
Redundancy and Contradiction, 53–54, 120
   check for, 75
   graphical, 102
Re-entering tables, 44
Reinwald, L. T., 173
Reinwald and Soland, the algorithms of, 174
Relational operator, 14
Relative Frequency: see Frequency
Repeat Character: see Ditto character
Replacements, table of, for V(B), 17
   for V(E), 18
Report Writer, COBOL, 66
Resistance to decision tables, 6
Restructuring tables, 111
RL–NR, rule number record, 88
Rules, 3, 7, 8, 13, 21
   addition of new, 109
   as-written, 58, 59
   checklist, 94
   check of, 99, 110
   combination of, 50, 56, 108
   combination, complex rules, 57
     extended entries, 58
     general requirements, 58
     independent rules, 58
     ranges, 58, 67
   complex, table for, 50; see also Rule, compound
   compound, 24, 58, 149
     simple rules in, 58
   continuation of, 86
   dependent, 54; see also Dependency
   ELSE-Rule: see ELSE
   hits, by transactions, 48
   hits, analysis of, 49
   horizontal in TABSOL, 4

   housekeeping, 44
   impossible, 57, 75
   independent, 54; see also Independency
   "practically possible," 56
   priority, 116
   selection of, 14, 28, 31, 32, 53, 110
     combining duplicate sections of, 50
     logical path length, 46
   simple, 19, 21, 55, 75
     conversion to complex rules, 57
   splitting of, 108
   uniqueness, 36, 53
Rule count, 58
Rule entry, 86
   boundaries, 86
Rule Mask, technique, 160
Rule Number, 86
Running Time, 45; see also Optimization, Runtime

SC, symbol for Split Count, 132
Scan, top-to-bottom, 41
Scanning procedure, technique, 160
Scientific Problems, 13
SECTION, COBOL, Decision table as, 88, 116
Semantics, 63, 64
Sentences, 39
   group of, 39
Sequential model of decision making, 6
Serial Execution, 19
Shorthand, 66
S, symbol for status of full set of n conditions, 14, 16, 17, 18
SIGPLAN, 5
Simple Rule Mask, technique, 162
Simple rules, 103
Size, of tables, 76, 80
SMP, 86
Software design, 74
Soland, R. M., 173
Source Language, minimization of decisions, 45
Source level debugging, 68
Special Interest Group for Programming Languages, 5
Split Count, 132
Splitting tables, 77, 132
Sprague, V. G., 172
Sprague's Counterexample, 172
Standard Format, decision tables, 51
Straight Scan, technique, 161
Structure of a decision table, 7–9, 40
Structures of tables, flowcharts, 116
Stubs, 7, 9, 86
   check of, 99
   condition, 9
   continuation of, 40, 86
Subordinate tables, 118
Subroutines, 44, 65, 66

closed, 91
Substitution Operator, and equivalent limited entry table, 35
Subtables, in decomposition, 48
Supporting theorem, 21
Sutherland Company, 3–5
Symbolic notation, 64
Syntax, 63
System, block diagram, 82
System Analysis, 4, 6, 73–84
    actions, 74
    calculations, 78
    causes, 78
    checklist, 93
    client, 74, 78
    commercial, 74
    conditions, 74
    effects, 78
    factoring of problems, 77
    interviewing, 74, 119
    logical organization, 78
    macro language for, 68
    post-interview analysis, 74
    steps of, 74
    tables, checking, 120
System Design Table, 79, 98
    analysis, 81
    division of, 82
    program flow analysis, 82
    transformation, 81
System Development considerations, 83
Systems, design of, 76

T, symbol for Table, 16, 18, 22
TABEND, table delimiter, 86
Table, checklist, 94
Table name, changing, 113
TABSOL, 3–5
TABTRAN, 86
Test Data, 77
    generated, 104, 105
Testing, 77

during production, 106
    see also Debugging
THEN, COBOL, 8, 64
Theorems, Decision Table, 21–24, 145–152
    I, 22, 145
    I', 22, 151
    I'', 22, 152
    II, 22, 146
    III, 23, 147
    III', 23
    IV, 23, 147
    IV', 151
    IV'', 24
    V, 24, 148
    VI, 24, 149
    VI', 24, 151
Theory of decision tables, 13–19
Time: see Running time
Time Dependence, 115
Top-to-bottom table analysis, 50; see also Decomposition, top-to-bottom
TRACE, COBOL, 106
    decision table, 104, 106
        simulation of, 106
Traditional Decision Tables, 28
    basic assumptions of, 28
    historical basis of, 54
Traditional Extended Entry, 29
Traditional Limited Entry, 28
Transaction, 28, 46, 53, 59, 78
Translation algorithms, 160–174; see also Decomposition, Algorithms
Translators, decision table, see individual names; Preprocessors
Trees, decomposition, 132
    of tables, 115
Truth Table, 3, 4
Turnaround time, 68

U, symbol for requirement, 18
United States of America Standards Institute, (USASI), 65
Unit-record card, an input record, 86

USA Standard COBOL, 154
User's Guides for preprocessors: see Manuals

V(B), symbol, 16, 17, 19
V(C), symbol, truth value of condition, 14, 15, 17, 18
V(E), symbol, 18
Variables, conditional, 17, 21, 157
    states of, 9
Verb, macro, 68
Verhelst, M., 164, 165

W, symbol, 16, 17
Worst case testing, 105
WRITE, COBOL, 66
Writing the decision table, 93–95
    primary restraint, 53

X: see Action marker

Y: see Yes
Yes, 8, 16, 18, 21, 28, 29, 30, 46, 50, 55

$: see Dollar sign
":", 35
..., 63
=. $\leqslant, \geqslant, <, >, \neq$: see Relational operators and names of operators
*: see Asterisk
-: see Don't-care
$\sim$, 16
+, 16, 18
".", 16
$\odot$, 19
$\oplus$, 19, 23
ϕ, Symbol for negation of I. 18, 21, 23, 24
ΔC, Symbol for Delta Count, 132
"$\rightarrow$", Symbol for implies, 19
2^n rules, 24, 56; see also Completeness
2x2 Method, technique, 165
88 items, COBOL, 77

T
57.4
P65

OCT 25 1971